MW01633912

GREENHOUSE GLASNOST

GREENHOUSE GLASNOST

The Crisis of
Global Warming

ESSAYS INTRODUCED BY
ROBERT REDFORD

TERRELL J. MINGER
editor

THE ECCO PRESS
INSTITUTE FOR RESOURCE MANAGEMENT
New York

The Ecco Press
26 West 17th Street
New York, NY 10011
Published simultaneously in Canada by
Penguin Books Canada Ltd., Ontario
Printed in the United States of America

Designed by Richard Oriolo

FIRST EDITION

The contributors and the IRM greatfully acknowledge the following sources:

"Innovative Policies for Sustainable Development in the 1990s" by Senators John Heinz and Timothy E. Wirth, and Robert N. Stavins is based on *Project 88* and a paper originally prepared for the World Resources Institute.

"Global Climate Change" by Thomas G. Lambrix originally appeared in the April 1990 issue of *World Climate Change Report.*

"Will Unexpectedly the Top Blow Off?" by Thomas E. Lovejoy is excerpted from an address given at the plenary session of the 39th annual meeting of the American Institute of Biological Sciences. It originally appeared in slightly different form in *BioScience* Vol. 38 #10, pp. 722–726. The author acknowledges the benefit of much discussion with Robert McC. Adams, William R. Graham, Martha Hays, J. P. Myers, Rob Peters, Peter H. Raven, and George M. Woodwell. Excerpts from "The End of the World" from *New and Collected Poems, 1917-1982* by Archibald MacLeish are reprinted by permission of Houghton Mifflin Co. Copyright © 1985 by the Estate of Archibald MacLeish.

"The Evolution of Gaia" by James E. Lovelock first appeared in slightly different form in *Nature* Vol. 344, pp. 100-102. Copyright © 1990 Macmillan Magazines Ltd.

"Croesus and Cassandra" by Carl Sagan is excerpted from his Oersted Medal Acceptance Address to the American Association of Physics Teachers in Alanta, January 1990, and will appear in the *American Journal of Physics* (on press). Copyright © by Carl Sagan.

"The Changing Climate" by Stephen H. Schneider first appeared in slightly different form in *Scientific American,* September 1989.

"Environmental Refugees" by Sir Crispin Tickell was first given as the Natural Environment Research Council Lecture at the Royal Society in London, June 5, 1989.

Library of Congress Cataloging-in-Publication Data

Greenhouse glasnost: the crisis of global warming:
essays/introduced by Robert Redford;
Terrell J. Minger, editor.
p. cm.
Includes bibliographical reference and index.
1. Global warming. 2. Global warming—Government policy.
3. Greenhouse effect, Atmospheric.
QC981.8.G56G74 1990   363.73'87—dc20   90-3432 CIP

ISBN 0-88001-258-7

The text of this book is set in Bodoni Book.

Greenhouse Glasnost is printed on acid-free recycled paper.

Acknowledgments

Each individual who contributed an essay to *Greenhouse Glasnost* is involved in some special way in making this world a cleaner, healthier, more sustainable place. And they are all the best at what they do. We are enormously grateful to each for the precious time spent in working and helping us with this project.

We would especially like to acknowledge academician Roald Sagdeev, a Soviet contributor and the head of the Soviet delegation to the Sundance Symposium. His global vision and practical support set in motion what we hope will be a long-term collaboration with colleagues and friends in the Soviet Union.

A very special thanks to Parry W. Burnap of IRM who, as assistant editor, did the real work, and to our editor at The Ecco Press, Lee Ann Chearneyi, without whom *Greenhouse Glasnost* would never have happened.

Though they were not directly involved with the publication of this book, we must acknowledge the faith and tireless work of the dedicated staff at the Institute: Paul Parker, Jon Lear, Connie Smith, Katherine Gigy, and Sharron Orton. They do the work of a group twice their size. They are responsible for the steps that led up to *Greenhouse Glasnost: The Crisis of Global Warming,* and they will be responsible for our future. We're also grateful for the financial support of the United Nations Environmental Programme, which helped make possible this valuable educational resource.

Finally, we need to highlight the support and encouragement of the founder of IRM, Robert Redford, and the patient assistance of Robbie Miller.

—T.J.M.

Greenhouse Glasnost is dedicated to the memory of
Dr. Walter Orr Roberts,
one of the world's true pioneers in the study of climate and man.
His work and life inspired this book and
the careers of many of our contributing authors.
We are all dedicated to carrying on his vision
of a world in which science serves the welfare of *all*
the inhabitants of this fragile planet, Earth.

CONTENTS

WALTER ORR ROBERTS

Preface:
Strategies for a Better Future in a Warmer World

Walter Orr Roberts, a distinguished astrophysicist, was founding director of the National Center for Atmospheric Research. He served as director from 1960 to 1968, and as president of the University Corporation for Atmospheric Research, which operates the NCAR, from 1960 to 1973. The purpose of the NCAR is to research weather, climate, air-quality chemistry, sun-earth influences, and the physics of the space environs of the sun. Roberts taught at Harvard University, the University of Colorado, and the Western Behavioral Sciences Institute. From 1974 to 1981, he directed the Program on Food, Climate, and the World's Future at the Aspen Institute for Humanistic Studies.

Early this year, Roberts succumbed to a struggle with cancer. His great talent and contribution to an understanding of the environment will be missed.

BACKGROUND

The possible future warming of the earth's climate as a result of the inadvertent release of additional greenhouse gases caused by human industry looms as a problem for future generations, of magnitude comparable with the threat of nuclear war. Not all of the effects will be adverse, but it is likely that most will be detrimental. It is the view of many experts on greenhouse warming that measures are needed to slow the onset of the probable warming and accommodate to those changes that cannot be averted. This book is a pioneering effort to identify the problem and devise strategies to deal with it. The book is distinctive in the breadth of its approach, covering as it does ethical issues, value considerations, religious thought, the role of the arts and the media, and politics and business, as well as scientific issues.

Some of the background for the Sundance "Summit" comes from the Greenhouse Glasnost Teleconference initiated by Roald Sagdeev, then director of the Space Research Institute of the USSR Academy of Sciences, Russell Schweickart, Apollo 9 astronaut, and me in 1987. We joined forces to establish an electronic mail teleconference with some twenty Soviet and American participants to consider a hypothetical scenario of a warmer world in 2050, in order to examine its implications for U.S.–Soviet relations and indeed for all the world. We wanted the group to examine social, political, economic, and ethical issues of a warmer world, but we did not wish to enter scientific controversies about whether this warming would indeed come.

We assumed:

- a warmer world, specified in a jointly derived scenario;
- a world population double that at present;
- an advanced information system and new technologies;
- a world free of the threat of nuclear war.

LIFE IN THE WARMER WORLD
OF 2050

Perhaps the strongest characterizing trait of the human race is adaptability. Most other forms of life on Earth live moment by moment far more at the mercy of chance vagaries of their physical environment. Through a couple of million years of existence Homo sapiens sapiens, our own subspecies of Homo sapiens, has begotten extraordinary skills for survival in the face of natural hazards. Only a few thousand years of this time span are chronicled in written records. The rest has been deduced from studies of ancient life forms through archaeological and paleontological research.

It's sad, of course, that in generating technologies for man's comfort in an often harsh natural world, we have created war tools of unparalleled destruction. As we develop new skills to improve the quality of life in a changing global environment, we must not forget the imperative to avert the ultimate catastrophe of an all-out nuclear conflagration. But we also need to remember that it will take equal ingenuity and wisdom to prosper in a warmer and more polluted earth of doubled population, which is virtually inevitable sometime in the next century.

As populations grow, and as industrial might reaches larger percentages of the burgeoning masses, we see the likely prospect of a global warming unprecedented since before the birth of mankind, perhaps even since extinction of the dinosaurs some 60 million years ago. This warming will be the product of the advancing industrialization brought about by human invention. What is the key to a good life in this not so distant future? It is, I think, to cultivate purposefully that characterizing trait of humans, adaptability. It strikes me especially that the United States and the Soviet Union must exercise leadership to avoid war and to help all nations invent a better future for mankind.

To enhance quality of life we must learn the secrets of adapting to an altered world. There are many ways to accomplish this, a few of which are suggested below. I urge readers of *Greenhouse Glasnost* to add items to the list and think about how to practice them.

• Attain more with less. Go farther on each tank of fuel used in transportation; do more conferring electronically instead of traveling to meetings; build homes warmed and cooled with less energy, using shade and sun rather than oil, gas, coal, and electricity where possible. These methods help slow the warming and give us more time to adapt.

• Control population growth. Further increase of greenhouse gases that produce the warming threat will depend in part on how many of the world's people are using energy and other products of modernization. If there are fewer of us, the problem, though still serious, will be reduced.

• Build forests on lands where the plow should not have broken the plains. Return marginal lands to perennial grasses; slow deforestation and replant harvested forests; encourage tree planting in urban centers.

• Use water more efficiently, and fertilizers, herbicides, and pesticides more judiciously; practice agroforestry where possible.

• Genetically engineer new species of food and fiber crops that thrive in warmer climates; seek out and enhance natural plants already adapted to warmer weather or drier growing seasons.

• Create institutions and infrastructures that demonstrate conservation, adaptation, and risk reduction in both the developing and the industrialized world; bring the more-with-less consciousness to the educational system from kindergarten through college.

• Pioneer novel technologies to do conventional tasks without using diminishing resources, as in seawater irrigation of food and fiber crops, which reduces the need for freshwater irrigation, as Carl Hodges and associates at the Environmental Research Laboratory in Arizona are doing; develop techniques to use conventional resources more efficiently, as in improved freshwater irrigation.

• Turn waste into new resources: recycle paper, metals, glass, certain plastics, and organic wastes.

• Internalize the costs of cleanup by implementing taxation of carbon dioxide and other gaseous waste and pollutant emissions; tax use of rivers and landfills for industrial wastes, sewage, domestic trash.

• Promote a stronger environmental ethic at all levels of public perception; elect officials committed to conservation of nature.

• Improve our scientific understanding of climate change and its impacts, so that whatever changes the global warming brings, we can take future-oriented adaptive measures soon enough for them to be maximally effective.

This is but a beginning. There must also be guarantees of access to food, shelter, and gainful work for the world's new billions. Severe toil and drudgery of existence must end for all humanity, leaving time and resources for improved education and training, cultural advances, and new creative endeavors. Global fairness must reach to the now poor third of the human race.

Many of the strategies to meet the aims listed above will bring benefits in a less polluted world even if there is no change in climate. Thus, uncertainties about future climate trends should not dissuade us from seeking them. Most will also provide still greater benefits if there is a hotter world ahead, as most climatologists expect. They are sound insurance for better life in a worst-case global climate-warming scenario. They are also guarantees for an enhanced quality of living in a best-case future climate scenario.

ROBERT REDFORD

Introduction

In the early 1970s, I was one of many working to prevent the construction of an enormous coal-fired power plant in southern Utah, in the spectacular and sacred four corners region. It seemed very simple at the time. I knew it would be a travesty to spoil that extraordinary land and foul that incomparably clean air for the purpose of feeding southern California's insatiable energy appetite.

We fought and won. Southern California Edison terminated its plans to build the plant. But to my amazement and hurt, I was burned in effigy by the local citizens. They had a different point of view. They wanted the jobs the plant would provide, and thought of me as an interloper.

That experience taught me an important lesson, one that led to the creation of the Institute for Resource Management and eventually to Greenhouse/Glasnost: The Sundance Symposium on Global Climate Change. I learned that there are legitimate differences of perspective surrounding environmental conflicts, and that successful resolution of those conflicts should not have to mean that one side has to be a winner and the other a loser. It is not a question of environmental quality versus economic development. It is a question of balancing what we need to develop for our survival and what we need to preserve for our survival.

After the 1970s, a decade that could withstand more radical gestures, the 1980s, with the advent of Reagan and Reagan policy, promised to be devastating for the environment. We were sure to run the risk of losing significant gains made in the late sixties and seventies. It would be a time when industry would flourish with favored treatment and unbridled approvals.

I founded the IRM in 1981 because it seemed to me there was a tremendous need to bring opposing sides together to solve environmental problems. In a curious way, at that time this was a more radical notion than just fighting brushfires. With problems of air, water, land, and energy increasing so rapidly, responsible resource management was necessary. The environmental movement's traditionally confrontational tools seemed to produce little more than marginal improvements in environmental quality—and in many cases appeared to actually delay progress by exacerbating a polarization of the issues.

Based on the idea that intelligent people will make the right decision if they are given all the information, the IRM established a process that focused on the human factor in environmental disputes. Its purpose was to find the common ground between traditional adversaries, the overriding force being the belief that such close, informal communication on a more personal level would lead to an increase in environmental sensitivity. We have worked on a wide variety of issues: resource development on Indian lands, the future of the electric utility industry, oil development in the outer continental shelf, management of U.S. Forest Service lands, and urban air quality.

In November 1987 in Denver, we assembled a group of environmentalists, business leaders, politicians, citizen activists, and public health officials to try to build a coalition and design a strategy to address local air-quality problems. During a lunch break I heard Dr. John Firor of the National Center for Atmospheric Science in Boulder present a slide show about the buildup of carbon dioxide in our atmosphere and the greenhouse effect. This talk fundamentally changed the way I looked at environmental problems and the role of the IRM. And it was instrumental in directing the work of the IRM toward what is now a heavy emphasis on global climate change.

We thought the IRM should apply this process to a large environmental problem before a stalemate occurred. In at least one instance, we wanted to move our search for common ground from local disputes to a global problem, and we thought that global climate change, although identified by many scientists as second only to nuclear holocaust in the threat it poses for mankind, was an issue still new enough that sides had not been polarized. Moreover, climate change clearly is a *global* issue that transcends all political and ideological boundaries.

Coincidentally, in May 1988 I was invited to the Soviet Union for a film retrospective, so we took advantage of the opportunity to approach the USSR Academy of Sciences on behalf of the IRM about collaborating on an environmental problem. I was invited to co-chair a workshop on global warming with the president of the Soviet Academy of Sciences. We called it Greenhouse Glasnost.

For me there were two discoveries. Soviet scientists, using different methods, were coming up with the same information as American scientists—that mankind is engaging in activity that will permanently alter the climate of the Earth. On a more basic level, I came to understand the Soviet people's deep feelings for their motherland.

As a result of the workshop, the IRM signed an agreement with the Soviet Academy of Sciences to hold Greenhouse/Glasnost: The Sundance Symposium on Global Climate Change in August 1989. The purpose would be to determine what aspects of global climate change the United States and the USSR agreed on and, after that, to identify some long-term opportunities to work together. We hoped our collaboration would send a signal to the world that the two superpowers most responsible for greenhouse-gas emissions were willing to take action together to solve the problems caused by climate change. Since we have set the example for pillaging the planet, it is time to set the example for preserving it.

Scientists have known about global warming for decades, but the rest of us have been slow to catch on, particularly our political leadership. A majority of credible scientists have reached a consensus. It is time to act. Climate is related to everything. The threat of global warming may force our attention, imagination, and resources and thereby be a catalyst for solutions to other major international problems: continued threat from expanding nuclear arsenals, the lack of stable international energy resources, atmospheric poisoning, the disparity between developing and developed countries, and population growth—to name a few. We have a chance to slow down, to turn the global climate crisis into an opportunity.

Before we can take advantage of this opportunity, I hope we can close the gap between scientific knowledge and public understanding. The purpose of our symposium—and of this book—is to go from computer models and databases to action, to pass the baton from the scientists who know what is going on to the rest of us who have to do something about it. The diversity of contributing authors here represents the mix of people who participated in the symposium, and it is intentional: scientists, artists, media experts, policymakers, environmentalists, business and industry leaders.

The greenhouse effect is a problem that is not merely scientific or political. It is perhaps too big to be solved by science or government alone. We all contribute to it and our children will feel its effects. We must all be part of its solution. Climate change is a human issue.

In the face of the crisis of global warming, we all become world citizens.

There is no limit to the human imagination, but there is a limit to our planet's resources. We need to learn more compassion, not just for our fellow human beings but for the fragile Earth and the other species with whom we share it.

Every generation has had its span of time and events to dazzle the mind. During my grandfather's span of time, he saw the changes from horses to cars to planes to jets to rockets! I am still witnessing the metamorphoses of the industrial age into the information age. Our children may be the first generation truly to wonder whether they are the beneficiaries of their parents' progress and development, or the victims.

We've always thought of climate as an act of God. It requires an enormous shift in the way we think of the world and our place in it to understand that we have already moved into an era in which we are actually responsible for managing climatic parameters. It's frightening, but it doesn't mean that we all have to sacrifice. Finally, after years of mistakes, we are coming to recognize that continued economic prosperity is tied to ecological stewardship. There is responsible profit to be made in caring for the planet.

It is my hope that *Greenhouse Glasnost* will play a small but important part as a catalyst for public understanding and concern about global climate change and its effects. We will be "solving" the climate crisis for decades to come. But only if everyone does his or her part.

As T. S. Eliot said, "There's only the trying, the rest is not our business."

—SUNDANCE, UTAH
July 1990

GREENHOUSE GLASNOST

CARL SAGAN

Croesus and Cassandra: Policy Response to Global Warming

Carl Sagan is the David Duncan Professor of astronomy and space sciences and director of the Laboratory for Planetary Studies at Cornell University. A Pulitzer Prize-winner, he is the author of more than 600 scientific papers and popular articles, and is the author, coauthor, or editor of more than twenty books including, *Cosmos*, on the *New York Times* best-seller list for 17 months. His Emmy and Peabody award-winning television series of the same name, has been seen in over sixty countries by more than 400 million people. The series included an early effort to alert the general public to the dangers of global environmental change.

Dr. Sagan has played a leading role in the Mariner, Viking, and Voyager planetary expeditions and for twelve years was editor in chief of *Icarus,* the principal professional journal in planetary sciences. His involvement in such environmental issues as greenhouse warming, ozone depletion, and nuclear winter arose from his work on planetary atmospheres: his doctoral thesis was on the Venus greenhouse effect. He has received the Honda Prize in Ecotechnology, the United Nations Environment Programme Medal, and the Leo Szilard Award for Physics in the Public Interest from the American Physical Society.

Apollo, an Olympian, was god of the sun. He was also in charge of matters other than sunlight, one of which was prophecy—that was one of his specialties. Now the Olympian gods could all see into the future a little, but Apollo was the only one who systematically offered this gift to humans. He established oracles, the most famous of them at Delphi, where he sanctified the priestess. She was called the Pythia. Kings and aristocrats—and occasionally ordinary people—would go to Delphi and beg to know what was to be.

Among the supplicants was Croesus, king of Lydia. We remember him in the saying "rich as Croesus." Part of the reason he was so rich is that he was one of the people who invented money—the first coins were Lydian, and were minted during Croesus's reign. (Lydia was in Anatolia, contemporary Turkey.) His ambition could not be contained within the boundaries of his small nation. And so, according to Herodotus's *History,* he got it into his head that it would be a good idea to invade and subdue Persia, the superpower of the seventh century B.C. Cyrus had united the Persians and the Medes and forged a mighty Persian empire. Naturally, Croesus had some degree of trepidation.

In order to judge the wisdom of invasion, he dispatched emissaries to consult the Delphic oracle. You can imagine them laden with opulent gifts—which, incidentally, were still on display in Delphi about a century later, in Herodotus's time. The question the emissaries put on Croesus's behalf was: "What will happen if Croesus makes war on Persia?"

Without hesitation, the Pythia answered, "He will destroy a mighty empire."

The gods are with us, thought Croesus, or words to that effect. "Time to invade."

Licking his chops and counting the satrapies shortly to be his, Croesus gathered his mercenary armies, invaded Persia—and was humiliatingly defeated. Not only was Lydian power destroyed, but he himself became, for the rest of his life, a pathetic functionary in the Persian court, offering little pieces of advice to often indifferent officials—a hanger-on ex-king. It's a little bit as if the Emperor Hirohito had lived out his days as a consultant on the Beltway in Washington, D.C.

Well, the injustice of it really got to him. After all, he had played by the rules. He had asked for advice from the Delphic oracle, he had paid handsomely, and it had done him wrong. So he sent another emissary to the oracle (with much more modest gifts this time, appropriate to his diminished circumstances) and asked, How could you do this to me? Here, from Herodotus's *History,* is the answer:

> The prophecy given by Apollo ran that if Croesus made war upon Persia, he would destroy a mighty empire. Now in the face of that, if he had been well-advised, he should have sent and inquired again, whether it was his own empire or that of Cyrus that was spoken of. But Croesus did not understand what was said, nor did he make question again. And so he has no one to blame but himself.

If the Delphic oracle were only a scam to fleece credulous kings, then of course it would have needed excuses to explain away the inevitable mistakes. Disguised ambiguities were its stock in trade. Nevertheless, the lesson here is germane: Even of oracles we must ask questions, intelligent questions—even when the oracles seem to tell us exactly what we wish to hear. Policymakers must not blindly accept; they must understand. And they must not let their own ambitions get in the way of understanding. The conversion of prophecy into policy must be done with care.

This advice is fully applicable to the modern oracles, scientists and think tanks, and universities. Policymakers send, sometimes reluctantly, to ask of the oracle, and the answer comes back. These days the oracles often volunteer their prophecies even when no one asks. In either case, policymakers must then decide what, if anything, to do in response. The first thing to do is understand. And because of the nature of the modern oracles and their prophecies, policymakers need—more than ever before—to understand science and technology.

But there's another story about Apollo and oracles, at least equally relevant. This is the story of Cassandra, Princess of Troy. It begins just before the Greeks invade Troy to start the Trojan War. She was the smartest and most beautiful of the daughters of King Priam. Apollo, constantly on the prowl for attractive humans (as were many of the Greek gods and goddesses), fell in love with her. Oddly—this almost never happens in Greek myth—she resisted his advances. She refused the overtures of a god. So he tried to bribe her. But what could he give her? She was already a princess. She was rich and beautiful. She was happy. Still, Apollo had a thing or two to offer. He promised her the gift of prophecy. The offer was irresistible. She agreed. Quid pro quo. Apollo did whatever it is that gods do to create seers, oracles, and prophets out of mere mortals. But then, scandalously, Cassandra reneged.

Apollo was not amused. But he couldn't withdraw the gift of prophecy, because, after all, he was a god. (Whatever else you might say about them, gods keep their promises.) Instead, he condemned her to a cruel and ingenious fate: that no one would believe her prophecies. (What I'm recounting here is largely in Aeschylus's play *Agamemnon.*) So Cassandra predicts to her own people the fall of Troy. Nobody pays attention. She predicts the death of the leading Greek invader, Agamemnon. Nobody pays attention. She even predicts her own early death, and still no one pays attention. They don't want to hear. They make fun of her. They call her—Greeks and Trojans both—"the lady of many sorrows." Today perhaps she would be dismissed as a "prophet of doom and gloom."

There's a nice moment in the play when she can't understand how it is that these urgent predictions of catastrophe—which perhaps, if her warnings were believed, could be prevented—are being ignored. She says to the Greeks, "How is it you don't understand me? Your tongue I know only too well." But the problem isn't her pronunciation of Greek. The Greeks' answer (I'm paraphrasing) is, "You see, it's like this: Even the Delphic oracle sometimes makes mistakes. Sometimes its prophecies are ambiguous. We can't be sure. And if we can't be sure, we're going to ignore it." That's the closest she gets to a substantive response.

The story is the same with the Trojans: "I prophesied to my countrymen," she says, "all their disasters." But they ignore her prophecies and are destroyed. Soon, so was she.

The resistance to dire prophecy that Cassandra experienced is equally stubborn today. Faced with an ominous prediction involving powerful forces that may not be readily influenced, we have a natural tendency to reject or ignore the prophecy. Mitigating or circumventing the danger might take time, effort, money, courage. It might require us to alter the priorities of our lives. And not every prediction of disaster, even among those made by scientists, is fulfilled. Most animal life in the oceans did not perish from insecticides; despite Ethiopia and the Sahel, worldwide famine was not a hallmark of the 1980s; supersonic aircraft do not threaten the ozone layer—although all these predictions had been made by serious scientists. So when faced with a new and uncomfortable prediction, we might be tempted to say: "Improbable." "Doom and Gloom." "We've never experienced anything remotely like it." "Trying to frighten everyone." "Bad for public morale." What's more, if the factors precipitating an anticipated catastrophe are long-standing, then the prediction itself is an indirect or unspoken rebuke. Why have we permitted this peril to develop? Shouldn't we have informed ourselves about it earlier? Don't we ourselves bear complicity, since we didn't take steps to ensure that government leaders took appropriate action? And since these are uncomfortable ruminations—that our own inattention and inaction may have put us and our loved ones in danger—there is a natural, if sometimes maladaptive, tendency to reject the whole business. It will need much better evidence, we say, before we can take it seriously. There is a temptation to minimize, dismiss, forget. Psychiatrists are fully aware of this temptation. They term it "denial." The rock group Dire Straits has a line in one of its songs: "Denial ain't just a river in Egypt."

The stories of Croesus and Cassandra represent the two extremes of policy response to predictions of deadly danger—Croesus himself representing the pole of credulous, uncritical acceptance, propelled by greed or other character flaws; and the Greek and Trojan response to Cassandra representing the pole of stolid, immobile rejection of the possibility of danger. The job of the policymaker is to steer a prudent course between these two shoals.

Suppose a group of scientists claims that a major environmental catastrophe is looming. Suppose further that what is required to prevent or mitigate the catastrophe is expensive: expensive in fiscal and intellectual resources, but also in challenging our way of thinking—that is, politically expensive. At what point do policymakers have to take the scientific prophets seriously? There are ways to assess the validity of the modern prophecies—because in the methods of science, there is an error-correcting procedure, a set of rules that have repeatedly worked well, sometimes called the scientific method. There are a number of tenets: arguments from authority carry little weight ("Because I said so" isn't good enough); quantitative prediction is an extremely good way to sift useful ideas from nonsense; the methods of analysis must yield other results fully consistent with what else we know about the universe; vigorous debate is a healthy sign; the same conclusions have to be drawn independently by competing scientific groups for an idea to be taken seriously; and so on. There are ways for policymakers to decide, to find a safe middle path between precipitate action and impassivity.

We sometimes hear about the "ocean" of air surrounding the Earth. But the thickness of most of the atmosphere—including all of it involved in the greenhouse effect—is only 0.1 percent of the diameter of the Earth. Even if we include the high stratosphere, the atmosphere isn't even 1 percent of the Earth's diameter. "Ocean" sounds massive, imperturbable. But the thickness of the air, compared to the size of the Earth, is something like the thickness of a coat of shellac on a schoolroom globe. Many astronauts have reported seeing that delicate, thin, blue aura at the horizon of the daylit hemisphere and immediately, unbidden, thinking about its fragility and vulnerability. They have reason to worry.

Today we face an absolutely new circumstance, unprecedented in all of human history. When we started out, hundreds of thousands of years ago, say, with a population density, averaged over the Earth, of a hundredth of a person per square kilometer, the triumphs of our technology were hand axes and fire; we were unable to make major changes in the global environment. The idea would never have occurred to us. We were too few and our powers too feeble. But as time went on, as technology improved, our numbers increased exponentially, and now here we are with an average of some ten people per square kilometer, our numbers concentrated in cities, and an awesome technological armory at hand—the powers of which we only incompletely understand and control. The inhibitions placed on the irresponsible use of this technology are weak, often halfhearted, and almost always, worldwide, subordinated to short-term national or corporate interest. We are now able, intentionally or inadvertently, to alter the global environment. Just how far along we are in working the various prophesied planetary catastrophes is still a matter of scholarly debate. But that we are able to do so is now beyond question.

Some of the gases in the air in front of us—carbon dioxide, water vapor, some oxides of nitrogen, methane, chlorofluorocarbons—happen to absorb strongly in the infrared, even though they are completely transparent in the visible. If you put a layer of this stuff above the surface of the Earth, the sunlight still gets in, but when the surface tries to radiate back to space, the way is impeded by the blanket of infrared absorbing gases. As a result the Earth has to warm up, to achieve that equilibrium between the sunlight coming in and the infrared radiation going out. If you calculate, from how much opacity these gases provide, how large the greenhouse effect ought to be, you come out with the right answer. You derive the correct, observed average value of the Earth's surface temperature. Without the greenhouse effect, that temperature would be well below freezing, the oceans would be ice, and we would all be dead.

Our lives depend on a delicate balance of invisible gases. A little green-

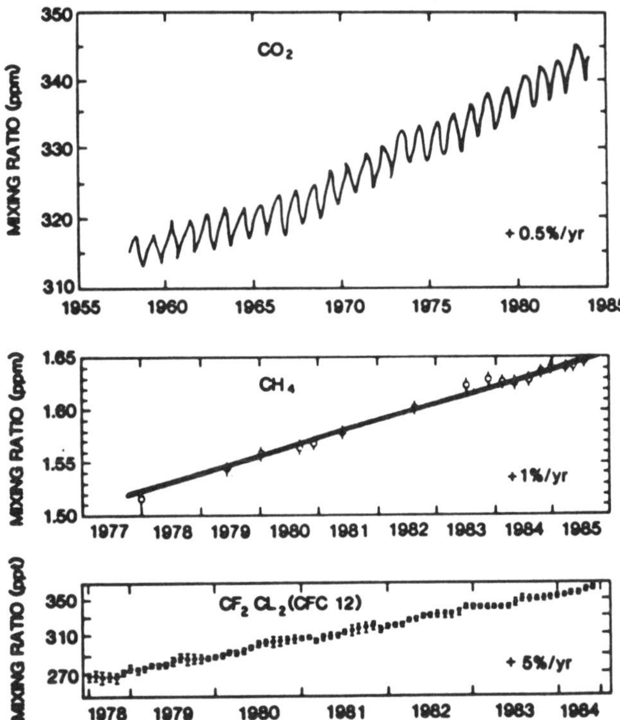

FIGURE 1. Increase during recent years of three of the principal greenhouse gases—carbon dioxide (CO$_2$), methane (CH$_4$), and a chlorofluorocarbon (CF$_2$Cl$_2$). The now classic CO$_2$ observations were obtained from the summit of Mauna Loa in Hawaii by C. D. Keeling.

house effect is a good thing. But if you add more greenhouse gases—as we have been doing since the beginning of the Industrial Revolution—you absorb more infrared radiation. You make that blanket thicker. You warm the Earth.

In Figure 1 is a representation of abundance, as time goes on, for three different greenhouse gases. Let's look at the first graph. The vertical axis shows how much carbon dioxide (CO$_2$) there is (in parts per million). The horizontal axis indicates the year (1955 at the extreme left). The top curve shows the increase over time of CO$_2$ in the Earth's atmosphere. The data come from the Mauna Loa atmospheric observatory in Hawaii. Hawaii is not highly industrialized, and forests are not burned here to the extent they are elsewhere. The increase in CO_2 with time detected over Hawaii comes from activities in other places on the Earth. The carbon dioxide is simply carried by the general circulation of the atmosphere worldwide—including over Hawaii. As you can see, every year there's a rise and fall of the gas. This is due to deciduous trees, which, in summer, when in leaf, take CO_2 out of the atmosphere, but in winter, when leafless, do not. Superposed on that annual oscillation is a long-term increasing trend, which is unambiguous. The CO_2 mixing ratio reached 300 parts per million in this century—higher than it's ever been during the tenure of humans on the Earth. Carbon dioxide absorbs in the infrared, we know its abundance is increasing, and this ought to produce an increase in temperature. The question is, How much?

Also represented in Figure 1 is the increase over time of methane (CH_4) and of one of the chlorofluorocarbons, the one with two fluorines and two chlorines; the increase for each is steep. As with CO_2, the level of CH_4 began growing with the Industrial Revolution. Chlorofluorocarbon (CFC) increases are the quickest—about 5 percent a year—because of the worldwide growth of the CFC industry. Note that CFCs are dangerous in two different ways: they attack the ozone layer, and they are greenhouse gases.

Various scientific groups—modern equivalents of the Delphic oracle— have calculated with computer models global temperature and precipitation changes. In Figure 2 is a comparison of the findings of five different groups working essentially independently with three-dimensional general circulation models, to predict by how much the temperature will increase if there is a doubling of the amount of carbon dioxide in the atmosphere, which there will be by the mid–twenty-first century. The modern-day oracles, listed in the left column, are the National Oceanographic and Atmospheric Administration's Geophysical Fluid Dynamics Laboratory at Princeton University; the National Aeronautics and Space Administration's Goddard Institute of Space Studies in New York; the National Center for Atmospheric Research in Boulder, Colorado; Oregon State University; and the United Kingdom Meteorological Office. In the middle column you can see that there is a difference of opinion by almost a factor of 2, but the average predicted temperature increase is something around 3° or 4°C (in Fahrenheit the numbers are about twice these). In the right column are predicted variations in the amount of precipitation. Again there are differences of opinion. Nobody claims that the predictions are perfect.

But note that none of the groups claims that doubling the carbon dioxide content of the atmosphere will cool the Earth. None of them claims that it

PREDICTING THE GREENHOUSE EFFECT WITH GLOBAL CLIMATE MODELS

MODEL	GLOBAL CHANGE IN:	
	TEMPERATURE (°C)	PRECIPITATION (%)
GFDL	4.0	8.7
GISS	4.2	11.0
NCAR	3.5	7.1
OSU	2.8	7.8
UKMO	5.2	15.8

RESULT OF DOUBLING ATMOSPHERIC CO₂ USING EQUILIBRIUM CLIMATE CONDITIONS

FIGURE 2. Projections for current state-of-the-art three-dimensional general circulation models of global temperature increase and precipitation decrease following a doubling of the atmospheric carbon dioxide mixing ratio.

will heat the Earth by tens or hundreds of degrees. We have an opportunity denied to many Greeks who consulted oracles—we can go to a number of oracles and compare prophecies. When we do so, they all say more or less the same thing. The answers are in fact in good accord with the most ancient oracles on the subject—the Nobel Prize–winning Swedish chemist Svante Arrhenius, who around the turn of the century made a similar prediction using, of course, much less sophisticated knowledge of the infrared absorption of carbon dioxide and the properties of the Earth's atmosphere. The physics used by all the groups mentioned above correctly predicts the present temperature of the Earth, as well as the greenhouse effects on other planets, such as Venus. Of course, there may be some simple error that everyone has missed. I'll say a word about possibly neglected factors shortly, but surely these concordant prophecies deserve to be taken very seriously.

We can take another approach to the problem. We can look at the actual record of global temperature change. In Figure 3 are the results from an ensemble of temperature-measuring stations all over the world; allowance is made here for the fact that cities are hotter than the countryside because of industrialization and higher population densities, as well as the fact that

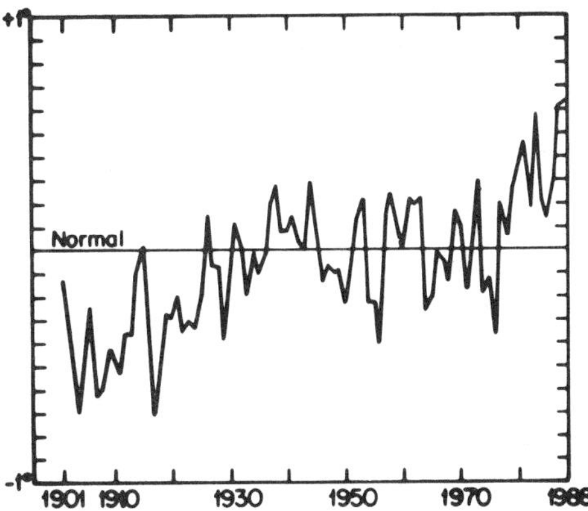

FIGURE 3. Mean global temperatures during the twentieth century. These temperatures have been found to be correlated with the CO_2 measurements of Figure 1 by the statistician Cynthia Kuo and her colleagues at Bell Laboratories. When allowance is made for the surface cooling in the early-to-middle 1980s from stratospheric particles injected by the El Chichon volcano, the correlation will very likely be improved. Data courtesy University of East Anglia and United Kingdom Meteorological Office.

cities tend to be darker and so absorb more sunlight. The temperatures shown range from the beginning of the century to 1988. You can see substantial wiggles, noise, in the global climatic signal. But there also seems to be a clear upward trend. The 1989 average global temperature, not represented here, is almost as large as the 1988 value. The five hottest years in the twentieth century occur in the 1980s. This sort of evidence has led some scientists to conclude that the signature of the increasing greenhouse effect is already here—not just something calculated for the twenty-first century, but here now. This is not to say that the drought of a particular summer was necessarily due to the greenhouse effect, but rather that the probability of such droughts and the probability of hot years is increasing as time goes on. It would be a very strange coincidence if the hottest years of the twentieth century occur just when the abundance of greenhouse gases is at its maximum, and the two events are not causally related.

Figure 4 presents a very broad perspective. At the left, it's a little more than 150,000 years ago; we have stone axes and are really pleased with ourselves for having domesticated fire. The global temperatures vary with time between deep ice ages and interglacial periods. The total amplitude of the fluctuations, from the coldest to the warmest, is about 5°C (almost 10°F). So the curve wiggles along, and after the end of the last ice age, we

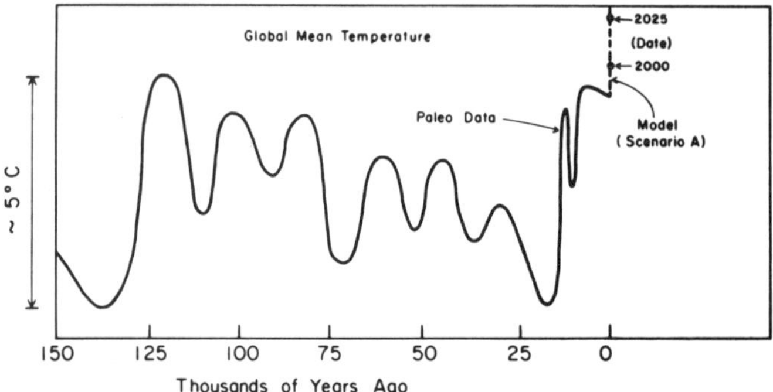

FIGURE 4. Evolution of global mean temperature over the last 150,000 years. Major glaciations (ice ages) are shown. Recent and projected near-future temperatures— from the intensifying greenhouse effect (significant efforts to reduce emissions of greenhouse gases to atmosphere are assumed)—are the warmest on record in this period. Schematic diagram courtesy James Hansen, NASA's Goddard Institute of Space Sciences (GISS).

have bows and arrows, domesticated animals, the origin of agriculture, sedentary life, metal weapons, cities, police forces, taxes, exponential population growth, the Industrial Revolution, and nuclear weapons (all that last part is invented just at the extreme right of the solid curve). Then we come to the present, the end of the solid line. The dashed line shows some projections of what we're in for because of greenhouse warming. This figure makes it quite clear that the temperatures we have now (or are to have shortly if present trends continue) are not just the warmest in the last *century,* but the warmest in the last *150,000 years.* That's another measure of the magnitude of the global changes we humans are generating, and their unprecedented nature.

Figure 5—from the work of NASA's Goddard Institute of Space Studies, directed by James Hansen—displays calculations of the average temperatures in four different years from the middle twentieth to the middle twenty-first century. It is much more difficult to calculate reliably the distribution of temperatures than it is the average global temperature. Nevertheless, these maps represent characteristic best estimates available at the present time. They assume that carbon dioxide, methane, chlorofluorocarbons, oxides of nitrogen, and other greenhouse gases will continue to be put into the atmosphere—but with some effort made to moderate their growth. The results are not encouraging. You can see the progression. At the top left is 1965 as a kind of baseline. Below it is 1990. At the top right is 2020, and at the bottom right is 2050. In 1965 and 1990, the dark areas

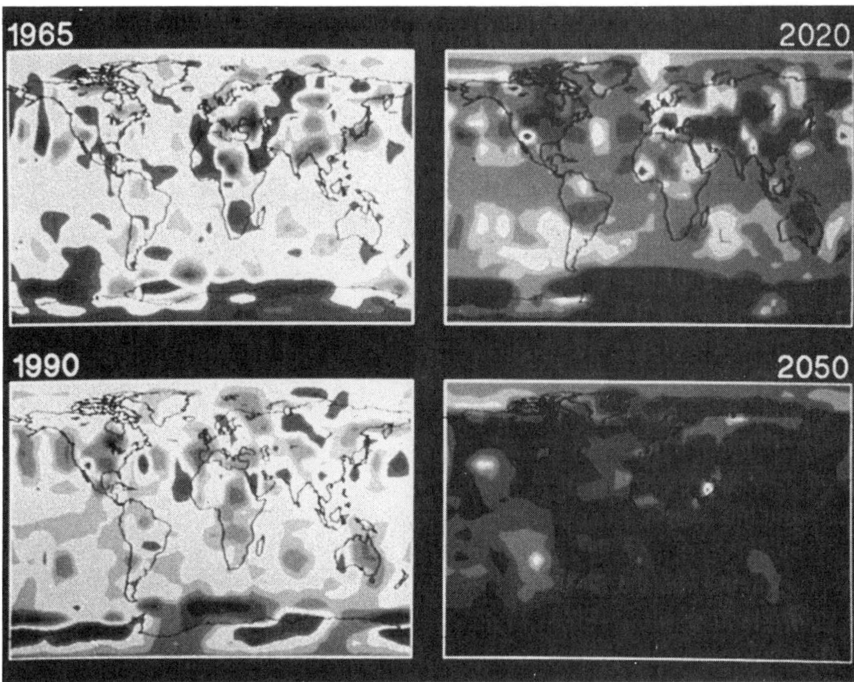

FIGURE 5. Global temperature changes from a three-dimensional general circulation model. Future projections assume a significant effort to mitigate emission of greenhouse gases to the atmosphere. Courtesy Dr. James Hansen, GISS.

are primarily areas of cooler temperatures and the greys show a slight warming trend. But in 2020 and, dramatically, in 2050, the dark areas and grey areas represent warmer temperatures. There is virtually no relief from the heat. Figure 6, for the same years, displays an index representing the probabilities of summer droughts. For this projection, which is far from the most extreme available, take a look at the American Midwest and the Soviet Ukraine—with much higher temperatures, much higher drought. Think of some of the agricultural consequences of those temperature increases. Notice that the climatic consequences are not restricted to one region. The warming is truly global. The drought is planet-wide. There are no places to escape to, no unaffected new locales for transplanted agriculture. In the long run, there are no winners from global warming. Everybody loses.

These temperature increases have secondary consequences that are also very serious. The temperature rise of a few degrees Centigrade by the middle of the twenty-first century is predicted to cause a thermal expansion of seawater and, to a lesser extent, a melting of polar and glacial ice. The

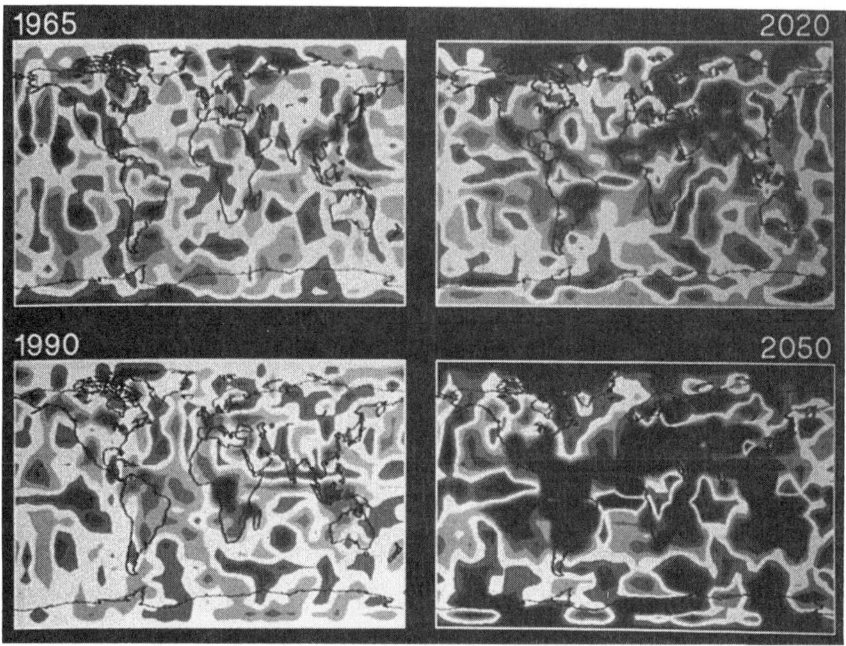

FIGURE 6. Summer drought index. Courtesy James Hansen, GISS.

consequence is a rise in sea level. Calculations suggest a worldwide average rise by a few tenths of a meter to a few meters by the second half of the twenty-first century. In the longer run, still more dire consequences may follow, including the collapse of the west Antarctic ice sheet and the inundation of almost all coastal cities on the planet. Heroic worldwide engineering would seem called for to prevent or even ameliorate the flooding.

Before I go on to explore further consequences of greenhouse warming and the nub issue of policy implications, let's consider some other aspects of validating the science. If you go to Antarctica or Greenland and you dig a deep boring core in the ice, the deeper you dig, the further back in time you are. The core doesn't go back very far in geological terms, but it's a very nice recent stratigraphic record. At a given depth, corresponding to a given time in the past, examine the ice. There are various indicators of what the temperatures were when that ice was laid down—from pollen studies (different plants proliferate at different temperatures), from radioisotope analysis, and in other ways. There are also trapped bubbles of air in the ice; the air can be extracted and the amount of carbon dioxide measured. There is a striking correlation: When the carbon dioxide abundance was high, the ancient temperatures were high; when the carbon dioxide abundance was low, the ancient temperatures were low. Clearly, this is consistent with the greenhouse story.

I promised I'd say something about feedbacks. This is one of the areas of uncertainty that policymakers must understand. There are both positive and negative feedbacks possible in the global climate system. Here's an example of a positive feedback: The temperature increases a little bit because of the increasing greenhouse effect and some polar ice melts. But polar ice is bright compared to the open sea. So as a result of that melting, the Earth is now a little darker; because the Earth is darker, it now absorbs more sunlight; it heats up more, it melts more polar ice, and so on. That's a positive feedback. Another positive feedback: A little more CO_2 in the air heats the surface of the Earth, including the oceans, a little bit. The now warmer oceans vaporize a little more water vapor into the atmosphere. Water vapor is also a greenhouse gas, so it holds in more heat and the temperature goes higher—a positive feedback, the dangerous kind.

Then there are negative feedbacks. They are homeostatic, like thermostats. Heat up the Earth a little by putting a little more CO_2, say, into the atmosphere. As before, this injects more water vapor into the atmosphere, but this generates more clouds. Clouds are bright, they reflect more sunlight into space, and therefore less sunlight is available to heat the Earth. The increase in temperature produces a decrease in temperature. Or put a little more carbon dioxide into the atmosphere. Plants generally like more carbon dioxide, so they grow faster, and in growing faster they take more carbon dioxide from the air—which in turn reduces the greenhouse effect. Negative feedbacks are like thermostats in the global climate. If, by luck, they were to be very powerful, maybe greenhouse warming would be self-limiting, and we could have the luxury of emulating Cassandra's listeners without sharing their fate.

The question is: Balance all of the positive and negative feedbacks and where do you wind up? The answer is: Nobody is absolutely sure. A recent estimate by Daniel Lashoff, then of the Environmental Protection Agency, is that the sum of all the positive and negative feedbacks could double or maybe even triple the temperature changes predicted for the next century that I've just summarized. Wallace Broecker of Columbia University points to the very quick warming that happened about 10,000 B.C., just before the invention of agriculture. It was so steep that he believes it implies an instability in the coupled ocean-atmosphere system; and that if you push the Earth's climate too hard in one direction or another, you cross a threshold, there's a kind of "bang," and the whole system runs away by itself to another stable state. He proposes that we may be teetering on just such an instability right now. Again, this consideration only makes things worse.

In any case, it's pretty clear that the faster the climate is changing, the more difficult it is for whatever homeostatic systems there are to catch up and stabilize. Also, I wonder if we're not more likely to miss unpleasant feedbacks than comforting ones. We're not smart enough to predict everything. That's certainly clear. I think it's unlikely that the sum of what we're too ignorant to figure out will save us. Maybe it will. But would we want to bet our lives on it?

Even cautious projections into the end of the next century imply major decreases in agricultural productivity because of drought connected with global warming. Forests and natural ecosystems are probably even more sensitive. Many nations are even today submarginal in their ability to feed their citizens. The decline of indigenous agriculture and the decreasing availability of imported food subsidies might be enough to produce a global famine. Drought in continental interiors would intensify; fresh water would become increasingly unavailable; desertification would accelerate; and, for variety, coastlines would be flooded. Some regions might get a little better; most regions, according to the projections, would get worse. You could say, "Okay, so the American Midwest will approach scrub desert, but then we can plant our wheat in Canada. What's the problem?" Well, even if the available unplanted topsoil in Canada were suitable for transplanted agriculture, and even if it really would be significantly cooler and wetter in Canada, moving all U.S. farming there is going to be enormously expensive. And the Canadians might have their own ideas on the matter and might not relish being flooded with this new variety of migrant farmer. But as Figures 5 and 6 suggest, Canada might not be much better off than the American Midwest. Also, it's much harder for forests to migrate poleward than crops. Pervasive and continuing drought has dire implications for hydroelectric power plants, dams, irrigation systems, and river diversion schemes. In the long run, there seem to be no winners in the era of global warming. It's not that some places are going to get better, others worse, and so on the whole if we are sufficiently detached, it's going to be a wash. The net result—if we wait long enough, and do nothing to prevent it—is predicted to be thoroughgoing, equitably distributed disaster.

Some scientists think the chance of a massive *agricultural* disaster from greenhouse warming by 2050 is low—perhaps only 10 percent. But the probability of a severe disaster need not be very high for us to take major, even heroic, steps to prevent it. The traditional military advice has been to take seriously not just the most likely actions of a potential adversary, but the worst of which it is capable. Trillions of dollars were spent during the Cold War as a result. Why should this prudent admonition be any less applicable to the world environmental crisis? And if 2050 is too soon, we need only wait another few decades. Unless we take mitigating steps, greenhouse warming will become progressively worse as the decades pass.

Consider a rise in sea level of about 1 meter. It would inundate many populated islands in Polynesia, Melanesia, and the Indian Ocean, and have devastating impacts on places as diverse as New Orleans, Venice, Bangkok, Miami, New York, and the Mississippi Delta. It would displace tens of millions of people in Bangladesh alone. There will be a vast new problem of environmental refugees. Where are they supposed to go?

Think about shared river basins, where two or more countries share a river. What happens when the water level gets dangerously low? Will there be water and food wars? Consider Japan, which imports more than half its food today; the Soviet Union, which depends on grain from North America (and Argentina and Australia). What happens to the Soviet Union if it continues not to be self-sufficient in agriculture and if grain imports dry up?

Think about how the global system works. People in the United States can't be bothered to drive fuel-efficient cars. They need gas-guzzlers, perhaps because cars are symbols—of status, freedom, manhood. So we put a lot more carbon dioxide into the atmosphere than would otherwise be the case, time passes, and then there are millions of refugees in Bangladesh and elsewhere who have nowhere to go. It's an interesting connection. It makes you think.

The biggest CO_2 emitter on the planet is the United States. Accordingly, it seems to me, the United States must bear a proportionate degree of responsibility for making sure the problem doesn't get worse. The next biggest CO_2 emitter is the Soviet Union. The third biggest, if we combine them, is all developing countries together. That's a very important fact; this isn't just a problem for the highly technological nations—through slash-and-burn agriculture, burning firewood, and so on, developing countries are making a major contribution to global warming. Next in order of complicity is Western Europe, then China, and only then Japan, one of the most fuel-efficient nations on the planet. Just as the cause of global warming is worldwide, any solution must also be worldwide.

The scale of change necessary to address this problem at its core is nearly daunting—especially for those policymakers who are interested mainly in doing things that will benefit them during their terms of office. If the required action to make things better could be subsumed in two-, four-, or six-year programs, politicians would be more supportive, because then the political benefits might accrue when it's time for reelection. But twenty-, forty-, or sixty-year programs, where the benefits accrue when the politicians are not only out of office but dead, seem less attractive to many of them. The benefits are in the future; the liabilities are now.

Certainly we must be careful not to rush off half-cocked like Croesus, only to discover that at huge expense we've done something unnecessary. But even more irresponsible is to ignore the problem and hope it will go away. Can't we find some middle ground of policy response that is appropriate to the seriousness of the problem but does not ruin us if somehow—through a negative-feedback deus ex machina, for example—we have overestimated the severity of the matter?

Say you're designing a bridge or a skyscraper. It's customary to build in, to demand, a tolerance to catastrophic failure far beyond what the likely stresses will be. Why? Because the consequences of the collapse of the skyscraper or the bridge are so serious you must be sure. You need very reliable guarantees of protection. The same approach, I think, must be adopted for local, regional, and global environmental problems. There is great resistance, in part because large amounts of money are required from government and industry. For this reason, we increasingly will see attempts to discredit global warming. But money is needed also to truss up bridges and reinforce skyscrapers. This is considered a normal part of the cost of building big. Designers and builders who cut corners and took no such precautions would not be considered prudent because they don't waste money on implausible contingencies. They would be considered criminals. There are laws to make sure bridges and skyscrapers don't fall down. Shouldn't we have similar laws and similar moral proscriptions for the potentially much more serious environmental issues?

I would like to offer some practical suggestions about dealing with climate change. I believe they represent the consensus of a large number of experts, although doubtless not all. They constitute only a beginning, only an attempt to mitigate the problem, but at an appropriate level of seriousness. To undo global warming and bring the Earth's climate back to, say, the 1960's will be much more difficult. The proposals are modest in another respect as well—they all have excellent reasons for being carried out, independent of global warming.

We need a quick phase-out of chlorofluorocarbons, which constitute about 25 percent of the greenhouse effect and are its most rapidly growing component. The 1987 Montreal Protocol calls for a 50 percent phaseout by the turn of the century. While a significant step forward, this is too leisurely a pace, and is likely to be transformed soon into a 100 percent phaseout by the year 2000. (The Protocol provides an important existence theorem: It is possible for major industrial states to take common action in the face of a serious global environmental threat. Perhaps we will see a similar protocol on CO_2 and other greenhouse gases before too long.)

We can require much higher energy efficiencies, and it's clear that it can be done with very little discomfort. Some of us remember Jimmy Carter shivering in his cardigan sweater in the White House. If that's what it takes to produce a world that is not significantly worse for our children and grandchildren than the one we were given, we should do it. But that's not even necessary. With tax-relief carrots and tax-penalty sticks, it should be perfectly possible to improve the efficiency of electrical power generation, make large-scale replacement of incandescent by fluorescent lamps, establish fuel-efficiency requirements for automobiles, and so on. (Automobiles are responsible for about a third of the carbon dioxide generated each year in the United States. Why do we tolerate cars that get only 20 or 30 miles per gallon, when it's entirely feasible to manufacture cars that get 60 to 100 miles per gallon?) In general, and especially in the long run, it is cheaper to conserve fossil fuels than to burn them; there are profits to be made in energy efficiency.

We must put much greater effort into renewable energy resources, especially solar energy. A small political drama in the White House, again during the Carter administration, related to this matter. A solar heating system was installed—not a solar-electric system, but one in which sunlight heats water and the hot water is used for various purposes, including presidential showers. The system's principal value, I suppose, was symbolic. The moment that Ronald Reagan became president, the solar heating system was ripped off the White House roof. The money allocated to research on renewable energy was cut drastically by the Reagan administration, and we marked time over the last ten years on this extremely sensible way to deal with the greenhouse issue. Meanwhile, as we have seen, all those greenhouse gases have been building up.

And yet solar power systems are nearly competitive with more traditional means of generating electricity, and if we include the environmental costs into the price charged per kilowatt-hour, solar power is probably already competitive. Nevertheless, much more research needs to be done on increasing the efficiency, and on transmission lines to carry power from the arrays of light-harvesting solar cells in sunny environments to more cloudy and more densely populated regions. Superconducting transmission lines need to be explored. Some studies suggest that this type of transmission may be workable even if we never get room-temperature superconductivity; transmission lines cooled even to liquid air temperatures may be economically feasible.

Other systems should of course be developed—especially wind, tidal, and hydrothermal technologies, which are renewable and nonpolluting. Systems that chemically burn hydrogen and leave only water as a waste product are possible and need further research and development. For a moment recently there seemed to be some prospect of low-temperature nuclear fusion, which under certain circumstances might have been the ideal solution to the combined greenhouse/energy problem. But that prospect has now rapidly faded, and we are left with hypothetical, enormous, expensive, high-technology fusion systems, which even their proponents do not imagine being available on a commercial scale for many decades. In any case it is hard to imagine such systems as the answer for the developing world.

That leaves fission nuclear power, which under some circumstances might be considered as the stopgap between the present reliance on fossil fuels and whatever technology that does not generate greenhouse gases will be adopted in the future. But as Three Mile Island and Chernobyl and the delinquency of many of the facilities supervised and run by the U.S. Department of Energy remind us, there is another price to pay if we adopt fission power. No greenhouse gases are generated, but serious accidents can happen (the total cost of Chernobyl is in the tens of billions of dollars). A deadly witch's brew of long-lived radioactive elements is generated that will burden our descendants for centuries and millennia to come. Especially after Chernobyl, worldwide public opinion seems dead set against major reliance on new fission power. If such power plants can be made safe and cost-effective, if there is an extremely reliable way to dispose of radioactive wastes, and if there is no way to divert fission products into making nuclear weapons, maybe fission power is the stopgap. But the burden of proof surely lies with the manufacturers and supporters of such facilities.

A large-scale worldwide research and development effort is urgently needed for the development of greater efficiency in the use of fossil fuels and of alternative energy sources economically and technologically appropriate for various regions of the world. Note, incidentally, that a mix of technologies will almost certainly be needed for a long time to come. High-intensity industrial power needs—for example, in steel foundries and aluminum smelters—are unlikely to be provided by windmills.

Then there is the question of forests. Trees eat carbon dioxide, removing it from the atmosphere. We need massive reforestation. Something like the area of the United States needs to be reforested to stabilize the global CO_2 budget at its present value in the face of anticipated emissions in the next century. What's happening today is the exact opposite. On the Earth at this moment, an acre of forest disappears every second. There are only so many acres on the planet. This is a matter, it seems to me, it is possible to do something about. Nations can plant trees. Individuals can plant trees. In recent years, Japanese companies have been the principal despoilers of tropical forests (especially hardwood), but there is a growing environmental consciousness in Japan. There are also nations, such as Brazil, that permit their forests to be cut down by foreign lumber companies (for a fee, of course) or permit their own citizens to cut down tropical forests to generate farmland (especially quixotic, because the tropical topsoil tends to erode quickly, in one or two growing seasons, after which there is nothing but scrub desert where once there was a great forest). But Brazilian opinion and policy is also swiftly changing. Perhaps such nations can be further encouraged in suitable ways—by better environmental education, say, or with reference to their international indebtedness—to reverse their course.

However, on such matters the industrialized countries do not arrive on the scene with clean hands. Admonitions from the United States to Brazil about environmental responsibility do not carry much weight unless this country acknowledges its abysmal environmental record and demonstrates major departures from business as usual. China, with the second largest coal reserves in the world, is not likely to deem credible any pleas from Western nations that China not industrialize as *they* did, unless those industrialized nations can demonstrate a willingness to make environmental sacrifices themselves—and perhaps also to provide the necessary alternative technologies. No nation is beyond reproach on global environmental issues. We must help one another.

Applied Energy Services, a company in Arlington, Virginia, has recently announced its intention to compensate for the carbon dioxide that will be produced over the lifetime of a coal-fired power plant it intends to build; it has arranged to plant new forests in Guatemala which will remove from the atmosphere the same amount of carbon dioxide its new facility will inject into the atmosphere. The intention seems fully meritorious, and I hope it will actually come to pass. This is model behavior. Shouldn't lumber companies be required to plant more forests—of the fast-growing, leafy variety, optimal for mitigating the greenhouse effect—than they cut down? What about the coal, oil, natural gas, petroleum, and automotive industries?

Why not a massive international commitment to replant forests?

There is one other matter central to global warming, especially in the long run, and that is world population growth. People make carbon dioxide. People have to stay warm. People have to cook food. People have to get to work. And these activities often involve putting greenhouse gases into the atmosphere. The more people, the more serious these problems are, and the more difficult it is to solve them. Curbing population growth is essential, and the way to do it is not just making contraceptives available and explaining family planning. Poor people are not too dumb to limit their family size, although that's a widespread misunderstanding in the West. People in developing countries have large families not only because of deep-seated human feelings but also as a kind of insurance, as a result of a cost/benefit analysis, because their governments do not provide social security. Children cost very little to raise, and will be useful in farming or whatever the meager family occupation is. People have many children because it is unlikely that most will survive to adulthood, and someone is needed to help with the work and look after the parents in their old age. All over the world there is an extremely interesting phenomenon known as the demographic transition. It works in communist and capitalist countries, in Buddhist and Christian countries, East and West; it's a transideological, transcultural reality. When the per capita income surpasses a certain level and people have enough to eat, the population growth rate drops suddenly and dramatically. The most effective way for the industrialized North to take care of rampant population growth is to help bring the billion poorest people on the planet to a degree of self-sufficiency. That's part of the solution to global warming as well.

Every one of the foregoing approaches to ameliorating global warming is desirable on quite separate grounds. Quick phaseout of CFC production is essential to safeguard the ozone layer. Higher energy efficiencies and the search for alternative, and especially renewable, energy sources are desirable for economic reasons, to minimize dependence on foreign oil and other energy sources, and as protection against the eventual depletion—whenever it comes (a controversial matter)—of fossil fuels. We are willing to allow environmentally risky off-shore oil drilling because it is important to preserve "energy independence." Doesn't this argument apply more forcefully to supporting research and development on alternatives to fossil fuels? Planting and preserving forests is important to preserve species diversity and for powerful emotional reasons: our pre-human ancestors used to live in trees, and so we retain a deep affinity for forests. Curbing global population growth is one of the imperatives of our time for reasons that are well known. Greenhouse warming is a policy issue that cuts across many of the other environmental and political issues of our time.

Global warming—as well as ozonosphere depletion, nuclear winter, acid rain, and many other of the new global environmental problems—has a new character. The key aspect is the irrelevance of national boundaries to the problem. Consider some country where fossil fuels are being burned. Carbon dioxide goes up into the atmosphere. Does it stay over that country? No. A CO_2 molecule doesn't know about political boundaries. It's never heard of passports. It has no brain, so it doesn't understand the profound idea of national sovereignty. It's just blown by the winds, moved by the general circulation of the planet. If it's produced in one place, it can wind up in any other place. The planet is a unit. No one nation can solve the greenhouse problem alone. Whatever their ideological and cultural differences, the nations of the world must work together; otherwise there will be no solution to greenhouse warming and other global environmental problems. We are all in this greenhouse together.

Notice also that there are no short-term solutions to this category of problems. When we put something into the air, it stays there for a long time. We have no way to flush it out. It may well be that what we do today will affect our children, our grandchildren, our great-grandchildren a hundred years and more into the future. If we mess up now, our children and their descendants will pay the consequences. These environmental issues force on us not just a transnational but also a transgenerational ethic. If we want to save ourselves, we are going to have to adopt—as Einstein said in 1945 in an only slightly different context—a new way of thinking. It will not be impossibly hard to do. Many major world religions have been teaching such doctrines for a long time.

I'm haunted by a sentence in the Book of Proverbs: "They set an ambush for their own lives." We cannot continue mindless growth in technology, with widespread negligence about the consequences of that technology. It is well within our power to curb technology, to direct it to the benefit of everyone on Earth. Perhaps there is a kind of silver lining to these global environmental problems, because they are forcing us, willy-nilly, no matter how reluctant we may be, into Einstein's new way of thinking. We are a resourceful species when push comes to shove. We know what to do. Out of the environmental crises of our time should come, unless we are much more foolish than I think we are, a binding up of the nations and the generations, and the end of our long childhood.

U.S. SENATOR
BILL BRADLEY

The Spirit of
Greenhouse Glasnost

Bill Bradley is currently serving his second term in the U.S. Senate, representing the State of New Jersey. Senator Bradley is a member of the Energy National Resources Committee and the Finance Committee, and he chairs the Energy Committee's Water and Power Subcommittee. He has written and sponsored a variety of environmental legislation and has initiated a number of efforts to preserve and protect New Jersey rivers, shores, and threatened wilderness. Senator Bradley also cosponsored legislation to ban dumping of sewage sludge into the ocean, wrote and passed a bill to prohibit the use of lead in new drinking-water systems, and sponsored legislation to monitor the disposal of medical waste.

Back in 1964, I was a member of the United States Olympic basketball team. I figured we would be playing the Soviet Union in the final game, so I went to a Russian professor at college and said, "Could you give me a few words I could use in case I get into trouble out there against the Soviet Union?" He gave me some words, and when the team got to the final game of the Olympics, we *were* playing the Soviet Union. I was my present height of six feet five inches, and weighed 195 pounds. My opponent on the Soviet team was six-foot-seven and weighed 240 pounds. About eight minutes into the game he cracked me with his elbow on my upper chest and lower neck, and I fell back. I remembered what the professor had told me, and I looked my opponent in the eye and said, *"Bud'te ostorozhno,"* which literally translated means "Hey, let's have a little mutual respect." And I hope that is the spirit of *Greenhouse Glasnost.* I want to salute Robert Redford for bringing together such an outstanding group of professionals from all walks of life, whose interest in this subject, the future of our planet, is deep and knowledgeable.

In July 1989, during the meetings of the Supreme Soviet, which were broadcast on television, something really extraordinary occurred. Prime Minister Nikolai Ryzhkov was nominating various ministers, who would hold various portfolios, and he nominated a man to be minister of forestry production. This man had passed through the committees and was recommended without reservation. He had addressed the Supreme Soviet, and the members were moving to a vote. Suddenly a deputy from Irkutsk, near Lake Baikal, asked to speak. He said that over a period of five to six years he had tried to reach the minister-designate about a pulp factory that was polluting a section of Lake Baikal, but he had failed to respond. Right after that, a man from a Baltic republic said: "We have a similar problem with a pulp factory, and this man was not responsive to us either." Soon after that someone else stood up and said, "We have had the same problem in the Ukraine." Five or six others stood up and said similar things. Still no one remarked that there was anything noteworthy in these speeches.

A vote was called for and subsequently taken. The results were handed up to General Secretary Mikhail Gorbachev, who was engaged in an animated conversation with Anatoliy I. Lukyanov, first deputy chairman of the

USSR Supreme Soviet. They looked at the card with the results, and Gorbachev seemed very surprised. He returned it to a clerk and said that something must be wrong; a recount was in order. The recount was taken. The card came back and the general secretary announced the results of one of the first votes of the Supreme Soviet on a nomination to one of the chief portfolios in the Soviet Union: 85 yeas, 70 abstentions, and 275 nays. The man was not confirmed minister, and the age of formal environmental activism in the Soviet Union was born.

Some people are surprised by that story. But I am not: I believe that one of the things Soviets and Americans share is a love of the land. For people of both nations that land is the wellspring of our greatness. It has steeled us, its beauty inspires our songs, its cruelty is the source of our sorrows. Early Americans were energized by the vastness of our territory and emboldened to start anew again and again on the frontier. They revered the land as the wellspring of strength and a root of values. They derived from the experience of the land a sense of independence, tempered by a respect for life, liberty, and the pursuit of happiness.

Unlike our brief American experience, Russian history goes back more than a thousand years. It is a history of triumph and tragedy, sometimes on a heroic scale. And it is always a history against the huge canvas that is the Soviet land. A majestic, silent procession of lakes and forests. The vast sweep of the steppes, the strong current of mighty rivers, the still sands of the arid zones, and the great Siberian wilderness of taiga and tundra.

Our peoples, American and Soviet, have been challenged and restrained by the land. We have trusted its generosity and too often have taken its replenishment for granted.

Now, from Love Canal to Chernobyl; from the befouled waters off Valdez, Alaska, to the desiccated salt flats around the Aral Sea; from the nuclear-waste dumps of Colorado and Ohio to the grimly polluted air of the steel and coal regions of the Urals and the Ukraine, we, all of us, see the land's vulnerability to abuse—and we recognize that its potential for giving us rebirth may be slipping away. The people of the Soviet Union and the people of the United States need the land not for the material wealth it provides but for a sense of who they are, both as individuals and as a people. We Americans are a diverse, independent, restless group. And most of us rarely catch a glimpse of unspoiled land. Our vistas are cluttered by billboards, neon signs, concrete. We leave our hometowns never to return. We change jobs as never before. We lose touch with each other. We marry, we split up, we remarry. We move faster and faster: instant coffee, fax mail, microwave meals. If we ever do slow down, it is because we are caught in a traffic jam or stacked up over an airport.

In contrast, the land offers a sense of permanence. Geologists call it "deep time." When you go, for instance, from the Grand Canyon in Arizona to Zion and Bryce national parks in southern Utah, you move up the stairs of hundreds of millions of years, recorded in the layers of the earth as if some ancient force wanted us to know its story and to remind us how, against that backdrop, the evening news, the musings of pundits, even our own lives seem but a nanosecond.

The land can be our teacher. It can instruct us in the values and the things that can't be bought or sold, traded or exchanged. And in our encounters with the land we can find freedom. Each of us can find something larger than we are, something that lasts longer than we do.

Four years ago, when I was in Irkutsk, I wanted to visit Baikal, the deepest lake in the world. I drove there in the afternoon after an exhausting all-night flight from Moscow. When I arrived at the lake, after all that distance, it was obscured in mist. I could see nothing, and my disappointment was as heavy as the fog. But early the next evening, after a long discussion with the great Siberian writer Valentin Rasputin, I went back to Baikal. This time the sky was clear and luminous, the lake stretched before us deep, still, pure. In nature's mysterious quietness I believed I could hear the heartbeat of time. I could sense the life-giving force that flows through all people, Soviet and American, who love the land.

Well, what about it? Will *perestroika, glasnost,* and democratization honor the land? Will the next forestry minister be better than the man who was rejected? It is really too early to tell. What needs to be done is clear. Begin effective reforestation. Reduce reliance on chemical fertilizers. Cancel big water projects. Clean up the water and the air (102 Soviet cities with population of more than 50,000 have air pollution levels ten times greater than the Soviet norm).

We know what to do. We know cleaning up the environment means not doing something else. The Soviet Union has a budget deficit three to four times greater than the United States in relative terms. That means that if we put environmental controls on existing plants and equipment, we don't build as much new industrial capacity. Investing in consumer production means we don't invest as much in sanitation or sewage. And adopting market pricing means that everything is more expensive, including environmental protection. Choices. That's what it's all about.

We all know that President Gorbachev and Foreign Minister Edward Shevardnadze have acknowledged the importance of cleaning up the environment. We all know that Article 20 in the 1987 Law Governing Enterprises has some very strong language on environmental cleanup and we know the ambitious goals that are being set for the year 2005. But as Americans, with the experience of the last several decades in a number of environmental areas, we might remark to our Soviet friends: To say the right things is admirable and easy; to do them, to commit resources and realize goals, is very hard.

As I contemplate our common environmental destiny, two things in particular occur to me. First, the image of the Delaware River, which separates my state of New Jersey from Pennsylvania. The same river that Washington crossed in the Revolutionary War—remember the painting? Twenty-five years ago, the Delaware was a dead river: no fish, no swimming. Then the Clean Water Act was passed. The act was enforced, and today the shad move 100 miles from the sea up the river. I swim in the river with my family, as do thousands of New Jerseyans—so it *can* be done.

The other thought I had concerns the common heritage and destiny of the United States and the Soviet Union. It is found in the words of the novelist Valentin Rasputin, who said about Lake Baikal what could be true about our Great Lakes, about all of nature: "Baikal was created not for production needs but for us to drink its water, marvel at its beauty, and breathe its precious air."

First and foremost, we *need* the Soviet and the American lands preserved for ourselves and our children—and our children's children. *Greenhouse Glasnost* is an early cooperative step toward translating that sentiment into a reality for Soviet citizens, for American citizens, indeed for all the people of the Earth.

GEORGII S. GOLITSYN

The Climate and Economic Priorities

Georgii Sergeyevich Golitsyn is a member of both the Institute of Atmospheric Physics of the USSR Academy of Sciences and the Presidium Academy of Sciences, and is an adjunct professor at the Moscow Physical Technical Institute. Academician Golitsyn's research is concerned with geophysical fluid dynamics, planetary atmospheres, nuclear winter, and climatology. He has collaborated with researchers from a number of countries including the United States.

CLIMATIC CHANGES

For the second year in a row, summertime rainfall in the European section of the USSR has exceeded the average standard rainfall, although a hot summer in 1988 followed a cold summer in 1987. In 1988, the United States experienced one of the worst droughts of the twentieth century. In Africa, extensive rainfall caused the Nile to overflow its banks for the first time in many years. Considered in isolation, all of these events could be explained as random climatic variations. They have always occurred, and they will happen again. But certain trends are emerging against these background fluctuations that suggest very serious changes on a global scale.

A striking example involves the Caspian Sea, on the USSR's southern border. We know the level of the Caspian Sea has varied by several tens of meters over the last million years. During the 1930s sea level dropped by 1.7 meters. It sank an additional 70 centimeters from 1930 to 1977. After 1977, however, the Caspian Sea began to rise. In 1989, sea level was approximately 1.5 meters higher than a decade ago. The press reported trouble along the beaches of Baku as well as flooding of commercial sites and other structures in the area. A powerful storm during the winter of 1986 washed out the railroad near Derbent; railway traffic was halted. The rising sea level and resulting elevated water table severely hindered construction in Makhachkala and other seaport cities. Eliminating the Caspian Sea's progressive water drop was paramount in the engineering and cost substantiation of a project begun in 1979 to redirect much of the runoff from northern rivers feeding the Volga; one has only to consider those circumstances, and the economic and ecological price for my country of accurate foresight concerning climate changes becomes obvious. This price may be incalculably large for the world.

Over the twentieth century, the average annual surface temperature on the Earth rose by half a degree Celcius, and with the unusually warm 1980s (1988 was warmer than any other year since measurements were begun in 1856) the planet has warmed by a total of 0.7°C. Such a rate of local temperature change is unknown in recorded Earth history.

Temperature, though, is not the only climatic element and is in fact not the most important element for human life. Precipitation is more important, particularly for agriculture. According to the laws of physics, a warmer atmosphere can contain a larger quantity of water vapor; the circulation of water in nature is accelerated (at least in the atmospheric loop) and the amount of precipitation increases on the average. However, the following question is of no small importance: How will this precipitation be distributed geographically and seasonally as the warming occurs? An analysis conducted by U.S., British, and Soviet scientists based on meteorological data from 1881 to the present has confirmed a certain rise in precipitation levels. The rise is most strongly evident in the central and upper latitudes, while a diminishing trend occurs in the subtropics, and in the tropics no precipitation changes have been seen at all. The rise in annual precipitation (of several percent) for the entire USSR is particularly evident. Seasonal levels of this precipitation and its distribution among the primary regions of the country remain subjects of future study. It is difficult to overestimate the importance of this research. The more timely and reliable it is, the more accurate our plans will be—to the benefit of both present and future generations. I wish to emphasize that the capacities of the Soviet national economy, as an integrated system encompassing the entire USSR, can substantially moderate the disturbing prospect of climatic changes for *isolated* regions but cannot eliminate the overall problem.

It will be important for scientists to correlate, with the utmost accuracy, seasonal precipitation distribution with the primary growing seasons of agricultural crops, and to determine the water content of soil in each specific region and how it changes with time. This process will depend greatly on the properties of the arable land and the soil-working techniques. (For example, non-terracing tillage substantially reduces moisture losses as well as soil erosion.) Water content of soil is measured at several agricultural meteorological stations around the USSR, but the spatial irregularity of the fields is so substantial and occurs on such a small scale that it remains unclear how representative the series of observations at these stations are for drawing any global, or even regional, conclusions; the situation is further exacerbated by crop rotation and changes in field-working methods. The difficulties of future research are great indeed, although efforts to surmount them have already begun in a number of scientific departments of the USSR State Committee on Hydrometeorology.

CAUSES

Very nearly all scientists share the belief that the primary cause of observed and anticipated climatic changes in the coming decades can be traced to alterations in atmospheric composition. One of the most outstanding achievements of modern geophysics has been the precise analysis of the Earth's atmospheric composition over the last 160,000 years. Cooperation between USSR and French scientists played a vital role in this discovery. The Soviet team drilled to record depths—up to 2 kilometers—in the Antarctic ice cap, while French scientists carefully analyzed the composition of air bubbles in the recovered core. (Ice in the lower core layers is 160,000 years old.) It has become clear that the two warm periods in Earth history—6,000 to 8,000 and 120,000 to 130,000 years ago—correspond to elevated carbon dioxide and methane gas concentrations in the atmosphere; in the ice ages, levels of these gases were approximately 1.5 times lower.

Carbon dioxide, methane, water vapor, oxides of nitrogen, and tropospheric ozone are all greenhouse gases. They freely transmit solar radiation to the terrestrial surface and absorb the thermal radiation of the planet and of the atmosphere itself. The greenhouse effect is responsible for a 32°C differential in the average Earth temperature (in comparison to our planet with no atmosphere at all).

It is clear from extensive recent research that methane concentration began to increase gradually in the late seventeenth century and afterward rose rapidly. (It is interesting that this closely follows the increase in the Earth's population.) Methane in the atmosphere has more than doubled over the last 300 years, and the current rate of growth, directly or indirectly related to human activity, is about 1 percent annually.

Carbon dioxide was first released into the atmosphere in large quantities in the late eighteenth century. The initial cause is deforestation because carbon dioxide is absorbed in photosynthesis. Emissions released from the combustion of fossil fuels—coal, oil, and gas—today represent the primary source of elevated carbon dioxide concentrations, which are now 25 percent higher than 200 years ago.

Nitrogen oxide concentrations rose by approximately 10 percent over the twentieth century. This increase can be attributed to the popularity of nitrogen fertilizers, as well as to the combustion of fossil fuels. The latter, by the way, has been a source also of the high concentration of sulphur oxides. These oxides of nitrogen and sulphur react with water vapor to form nitric and sulphuric acids—creating the acid rain that inhibits and kills forests, plankton, fish, and so on. Obviously, changes in the composition of the atmosphere are directly related to environmental damage.

Chlorofluorocarbons (CFCs) are very dangerous. These compounds of fluorine and chlorine with carbon are used in the refrigeration industry and aerosol cans. If there were no CFCs in the early 1950s (these are not natural gases but are exclusively products of human activity), the concentration of CFCs in the air has risen to exceedingly high levels today (from 5 to 10 percent each year). CFCs decompose at very slow rates under natural conditions (tens or hundreds of years), which accounts for their rapid accumulation.

Chlorofluorocarbons have been linked to the ozone "hole" discovered in 1985 over Antarctica. Almost fifteen years earlier, U.S. scientists had warned that their manufacture would lead to a breakdown of the ozone layer, although it was expected the first indicators of this dangerous phenomenon would appear only by the year 2000. Nevertheless, the U.S. government banned the use of aerosol cans with CFCs in the seventies. A few countries, including Sweden, followed this example. But the move was certainly not enough, and the level of chlorofluorocarbons in the atmosphere continued to rise.

Most recently, wandering "holes" have been sighted in winter and springtime at upper and even moderate latitudes in the Northern Hemisphere. A thorough analysis of satellite and ground data reveals that the total concentration of atmospheric ozone dropped by 3 percent in the 1980s, primarily as a result of increasing concentrations of CFCs.

How are we to evaluate the effect of these gases on climatic changes? This critical scientific problem requires the coordinated efforts of climatologists and meteorologists, oceanologists and geologists, physicists and chemists, mathematicians and biologists, hydrologists and glaciologists, space researchers, and many others. Around the globe, work continues within the framework of the World Climate Program, the developing International Geosphere-Biosphere Program Global Changes, and a number of other programs sponsored by the World Meteorological Organization and the International Council of Scientific Unions, which include as members the scientific academies of many nations. All of these programs are designed to run beyond the year 2000. The USSR does participate in these programs, although I believe its contribution could be much greater.

A quantitative measure of the greenhouse effect is the thermal radiation trapped in lower atmospheric layers that heats the terrestrial surface. The additional rise in this radiation over the last 200 years, due to the increasing concentration of greenhouse gases, has been estimated at about 2.5 watts per square meter. Approximately 45 percent of the rise in the thermal radiation comes from carbon dioxide, 23 percent from methane, 19 percent from CFCs, and 3 percent from nitrogen oxides. These figures may be refined and their ratios may change in the future, as scientists make more accurate calculations and as reforms are initiated in worldwide industrial manufacturing. A possible reduction in CFC output, for example, may result from the 1987 Montreal Protocol, which was added to the "Convention to Protect the Ozone Layer"; the USSR has adhered to the Protocol since the fall of 1988.

Physical mathematical modeling is the most popular method of estimating future climatic changes. Modeling techniques are used to describe the conversions of solar and thermal energy and their interaction with clouds, various oceanic and terrestrial processes, winds, water transport, and so on. The ability to classify these processes in both space and time—as well as the range of their description—depends on the capacities of modern computers. Unfortunately, computers are ultimately inadequate, particularly in dealing with oceanic phenomena. However, most of the processes themselves, such as the interaction of solar and thermal radiation with cloud cover, and the interaction of the atmosphere with the underlying surface (ocean, terrestrial surface, ice cap), have yet to be fully studied, and present knowledge of these phenomena is not an entirely suitable basis for mathematical models.

Many estimates suggest that a doubling of the carbon dioxide concentration in the atmosphere will cause an average temperature increase at the Earth's surface of between 1.5° and 4.5°C. Why such a range in the predictions? Insufficient knowledge of these processes.

Climatic modeling suggests that against a background of overall warming, winter temperatures are increasing more rapidly than summer temperatures, and the higher the latitude, the stronger the effect. This is generally consistent with meteorological data for the past century. Analogous pictures emerge from analyses of the paleoclimate of warm eras: some 6,000 to 8,000 years ago, the average global temperature was approximately 1°C higher than the present temperature, while in the preceding interglacial period, 120,000 to 130,000 years ago, it was approximately 2°C warmer.

POSSIBLE CONSEQUENCES

The crude nature of modern models makes their predictions vary substantially for specific regions. Therefore, in their practical recommendations Soviet scientists rely largely on paleoclimatic reconstructions to estimate climatic changes. It is important that the general structure of a warmer climate be identical in predictions based on models *and* in those based on reconstructions. Experts claim that climatic changes could be favorable for the USSR, at least agriculturally. But at the same time, an increase in precipitation will cause more frequent flooding—as we've already seen, the rising level of the Caspian Sea presents a serious problem. The location and frequency of droughts remain unclear. To put it simply, the specific details of the consequences of climatic changes are yet to be determined; no one has begun such a serious study. Meanwhile, time is wasting. . . .

There will be serious consequences for the rest of the world as well. Paramount among these is a further drying out of the subtropics and certain areas of the tropics. Consider the Sahel, a region to the south of the Sahara. Hundreds of thousands of people have starved to death there in recent years as a direct result of drought. Millions of Africans have migrated. In the summer, parts of the United States could become desertlike, and this in a nation that accounts for approximately one-third of imported agricultural goods to the USSR. Where will this lead? In recent years the International Council of Scientific Unions and the United Nations have sponsored broad-ranging research into the possible consequences of nuclear war. One important conclusion of these studies centers on crop losses from "nuclear winter": serious disruptions to the international agricultural trade will cause 1 to 4 billion deaths from starvation. Extrapolating, we can assert that the threat of hunger is now one of the primary dangers of climatic changes. The agricultural balance of many developing nations, particularly in Africa, depends largely on the importation of cheap produce, primarily from the United States. Any change in this balance, any repudiation of single-crop economies (coffee, cocoa) to adapt to new climatic conditions poses difficult problems—of a social, political nature and, more basically, of plain survival—requiring many years of careful adjustment.

A rise in the world's seas and oceans is a dangerous consequence of global warming. Thorough research begun in the USSR by Professor R. Klienge and continued in the United States suggests that ocean levels have risen by approximately 10 to 15 centimeters during this century. To some degree, this can be attributed to the water expansion that occurs when water is heated, to thawing glaciers. Certain estimates claim that the level of seas and oceans may rise by as much as 1 meter by the mid–twenty-first century. These estimates are especially relevant when one considers that approximately 30 percent of the Earth's population lives in coastal regions within 50 kilometers of the water. Although the process extends over many decades, losses resulting from the rise in water level will represent a serious threat to the overall world economy.

Permafrost thawing is a significant, albeit regional, consequence of climatic changes. It will be felt most strongly in the populated regions of northern territories, where the social and economic infrastructure (roads, buildings, communications) will suffer substantial damage. Most of the thawing today comes from direct human intervention in the highly delicate and sensitive natural processes of polar territories. However, scientists have already identified a natural trend toward permafrost thawing, and this problem has become increasingly urgent.

The broad variety of consequences from climatic changes for the entire socioeconomic sphere are so extensive they can no longer be ignored. In December 1987, General Secretary Mikhail Gorbachev and President Ronald Reagan signed a joint declaration in Washington, D.C., naming climatic change as one of the most important problems facing mankind. American and Soviet scientists were asked to develop scenarios of climatic changes through the mid–twenty-first century. This joint effort should be completed by spring 1990.

The gloomy prospect of environmental damage caused by climatic warming, the reduction in precipitation in many areas coupled with rising ocean levels, looms before the whole world. Mankind must ponder these questions: Is it right for us to leave our children and grandchildren a world that has fewer resources and is less healthy than the world we inherited from our parents? Is there a way out of a situation whose natural development threatens to end in a genuine catastrophe?

WHAT IS TO BE DONE?

All of these issues have long been discussed by scientists. They have been taken up by Western social organizations such as the so-called Greens, who today have acquired a widespread and critically important sociopolitical voice. The International Environmental and Development Commission, or Brundtland Commission (named after its chairman, the then prime minister of Norway), was founded under United Nations sponsorship in 1983 and included representatives from twenty-one nations (the USSR delegation included academician Vladimir Sokolov). The commission completed its work in late fall 1987, and its report, *Our Common Future,* was published in early 1988. The book is extraordinary for the breadth of its coverage of the issues and its treatment of all aspects of socioeconomic life. The commission concluded—to put it briefly—that mankind has a promising future if it can bring to an end the thoughtless extensive development responsible for the waste of natural resources, for rising levels of environmental pollution, and for the broadening rift in living standards between rich north and poor south.

How can this be done? It can be done through substantial improvement in the efficient use of energy and the natural resources that produce it, through the reclamation and recycling of so-called wastes, through the sharing of energy- and resource-saving technologies with countries that don't have access to them and are therefore forced to manufacture products by means of wasteful techniques. According to Western experts, investment in energy- and resource-saving technologies normally has a payback period of several years, occasionally as few as one or two. To modify industry to incorporate comprehensive energy and resource savings not only will solve, or at least moderate, environmental problems including those that involve climate, but also will be advantageous from a purely economic standpoint and will diminish the gulf between generations. Parents have always believed their sons and daughters will have a better life than their own. This belief is in jeopardy now. To restore it we must rebuild the worldwide economy, and the socioeconomic consciousness of all mankind.

Why the worldwide economy and the socioeconomic consciousness of all mankind? Because we share one Earth, one common atmosphere, and a single ocean. The belief that the atmosphere, the ocean, the entire planet are the common property of mankind is becoming increasingly popular—as is the belief that we must protect them from ourselves and from our present economic policy. The profligate waste of natural resources and the widening inequality between rich and poor nations are intolerable, enormous problems.

I will cite only one possible route for slowing the rate of change in the composition of atmospheric gases and thus of the greenhouse effect. I propose taxing every ton of fossil fuel consumed in developed nations at a small percentage of its cost. From these proceeds an international assistance fund could be established for developing nations to introduce energy- and resource-saving technologies and end deforestation—particularly in the tropical rain forests, which, in absorbing carbon dioxide from the atmosphere and producing oxygen, are the world's lungs.

The ideas that mankind's development need destroy neither the environment nor quality of life and that global problems are intimately interrelated are entering the consciousness of society in general and government institutions. The consistent policy of peace advocated by the USSR and many strong Soviet initiatives have produced an objective dynamic whereby humanity can believe quite seriously that nuclear war is avoidable. Quality of life, then, becomes our focus.

The year 1988 was critical in this respect. It was then that the European community (including twenty-one West European countries), followed by the Group of Seven (Canada, the Federal Republic of Germany, France, Great Britain, Italy, Japan, and the United States), supported the Brundtland Commission's conclusions and proclaimed as sustainable its goal, a form of socioeconomic development that would not exhaust natural resources or the environment. These conclusions found additional support in a document entitled "The Consequences of the Arms Race for the Environment, and Other Aspects of Ecological Security," issued by the Political Consultative Committee of the Warsaw Pact nations. This document states that "economic activity without regard to ecological factors will lead to pollution of the seas, oceans, and atmosphere, unlimited propagation of pollutants, soil degradation, desertification and deforestation, climatic transformations, destruction of entire species, accumulation of toxins in the biosphere, and many other negative phenomena in nature that disrupt the human habitat. . . . Life itself urgently requires pooling of the efforts of the world community in the interest of a joint and effective solution to environmental problems."

Hence an immediate restructuring of the entire sociopolitical consciousness is needed. Characteristically, Western scientists, social figures, and even representatives of government circles relate this restructuring process to the current policy of renewal in the USSR and Eastern Europe. Enormous effort, however, is required to implement initiatives for change. I will consider briefly the present potential of the world for progress.

Approximately 22 billion tons of carbon dioxide were released into the atmosphere in 1987. Of this amount, 45 percent can be attributed to industrial sectors (metallurgy, chemistry, paper and construction-materials production). The amount dropped by 6 percent between 1973 and 1981 in the United States, at the same time that production levels rose by 13 percent. During this same period Italy reduced consumption per unit of product by nearly 6 percent annually. Japan has similar figures. While this savings trend has slowed in the Western nations in recent years because of the sharp drop in oil prices, scientific research and industrial innovations have accelerated. The successes here are simply astounding, and provide some hope for a global solution.

U.S. scientific research reveals that it is now possible to reduce electricity costs for lighting by a factor of 4 to 10, for water heating by a factor of 4 to 7, for domestic home heating by a factor of 6 to 10, and for refrigerators by a factor of 6 to 7. This can all be achieved without diminishing quality of life. According to United Nations forecasts, the savings in energy-producing fuel resources earned through the introduction of new energy-efficient technologies will, by the year 2000, reduce energy consumption 19 percent in Western Europe, 22 percent in the United States, and only 16 percent in the USSR.

In the worldwide market, the introduction of new economical (yet expensive!) devices normally requires intervention of government to establish specific standards. There have been many examples in the last ten years in the United States itself, however, in which companies that produce electricity have provided subsidies to consumers who purchase more economical appliances, since it is more advantageous to sell less energy than to construct new power plants. Conservation is often much more profitable than expanded production.

WHAT CAN WE DO?

Why do we have such poor energy-use figures? In a poorly organized carbon combustion, 40 percent is from oil burning, and 15 percent from natural gas. Oil yields 15 percent, and gas 43 percent, less carbon dioxide than coal for energy-output. Simply by substituting certain types of fuels with others, it is possible to reduce significantly the volume of carbon dioxide released into the atmosphere. As we know, the reserves of combustible fuel, particularly oil, are limited and must be carefully conserved for future generations. There may also be a better application for oil—petrochemistry, for example. Instead of burning the oil, chemicals derived from it through petrochemistry could be used to manufacture renewable goods.

How are these 22 billion tons distributed among the world's nations? The greatest percentage comes from the United States, 23 percent, followed by the USSR with 19 percent, Western Europe with 13.5 percent, China with 8.7 percent, and Eastern Europe with 7 percent. The remaining nations of the world account for a little less than 29 percent.

Energy efficiency, calculated to produce a unit of Gross National Product (GNP), varies substantially from country to country. France has the best record here, most likely because of its well-developed service sector. If we use France as a reference point of 1, the energy efficiency figure is 1.13 for Japan, 1.37 for the Federal Republic of Germany, 2.0 for Great Britain, 2.24 for the United States, 3.13 for Poland, 3.76 for the USSR, and 4.75 for China. My country uses nearly four times as much energy per unit of GNP as France and approximately 70 percent more than the United States—whose figure already leaves much to be desired (and this creates no small concern in the American government and among business leaders and the public at large).

Especially today, there is enormous potential for saving energy in the world. Western nations initially addressed this problem by enacting more energy-efficiency measures in the mid–1970s after the oil embargo and sharp rises in oil prices. Greedy energy consumption by the main economy that gives rise to a bureaucratic mentality and corrupt policies puts powerful brakes on scientific and technological innovation. Industry and modern agriculture are the primary environmental polluters. They are also the primary sources of greenhouse gases. Today, the USSR accounts for 20 percent of the world's oil and steel production, 25 percent of the world's mineral fertilizer production, and 40 percent of the world's natural gas consumption, while the nation's deforestation level is comparable to that of the United States and its production is two to three times lower.

Here is just one example of the bureaucratic mentality. The USSR long ago developed low-pressure turbine designs that employ the heat produced in the production of electricity. What an efficient arrangement! But the manufacture of these low-pressure turbines has been bogged down in disputes between many ministries: Who is to make them?

The agricultural dilemma was acknowledged at the March 1989 Central Committee Plenum as one of the most serious problems in the USSR. How to solve it? According to official data, up to 20 percent of produce in short supply in the USSR is lost in the storage of finished agricultural products. I would expect these figures to be much greater for vegetables and fruit; this can be determined by visiting vegetable stands or reading the newspapers every day. The people call a vegetable storehouse a vegetable rothouse with good reason.

The Soviet Union purchases between 25 and 45 million tons of grain annually—and probably loses a similar amount, since the grain is stored under unsuitable conditions. At an annual harvest of approximately 200 million tons (averaged over the last Five Year Plan) total elevator capacity proved insufficient: only about one-half of the harvest was stored there. The remaining grain was kept at collective and state farms, where it either rotted or was burned. At the same time, the United States, which has an annual grain harvest of approximately 300 million tons, has elevator capacity for 600 million tons. This makes it possible (even with substantial exports) to store the product for more than a year and to guarantee proper storage in bumper-crop years.

Unfortunately, the Soviet State Committee for Agriculture has been designed for average harvests—in bumper-crop years, losses have been simply staggering. The course toward new system of land ownership, a restructuring of agriculture, will spur a growth in production of agricultural goods that cannot, at current capacities, be stored anywhere. In order to avoid complete spoilage of new harvests, it is necessary to alter current investment policies radically, to invest capital in mechanical improvements for agriculture, to develop the infrastructure of the agroindustrial complex—roads, elevators, warehouses, storage facilities—and to bolster agricultural distribution. A number of proposals have been made in recent years, although so far their implementation is not progressing well.

New arable lands and the Volga–Don and Volga–Chogray channels, for which the Ministry of Reclamation and Water Management has continuously fought, are not needed. Economic investment in the agroindustrial infrastructure has demonstrated itself to be much more advantageous and has reached the payback point much more rapidly than have large water treatment facilities (which cause environmental damage). It is again convenient to refer to U.S. experience. The federal government decided in 1976 to end supports for such projects, with the exception of water reclamation and economy plans. Professor Gilbert White, a prominent American geographer and specialist on water and environmental issues and a foreign member of the USSR Academy of Sciences, in discussing the developmental strategy of water use around the world, emphasizes that regardless of how the climate changes there can be only one water-use strategy: uniform water conservation, because this is, above all, cost-effective. He even neglects to cite environmental conservation, since he considers it to be self-evident. Such an attitude in the United States has resulted in a reduction in water consumption of 10 percent between 1980 and 1985, with an approximate increase of 15 percent in the Gross National Product.

The restructuring of our USSR economy is intimately related to the environment, including climatic changes. This relationship is becoming the focus of increasing attention around the world. The USSR is developing a course of energy and resource conservation combined with accelerated production. But it will be necessary to solve certain ecological problems to guarantee that it will not stray from the goal.

It is impossible to overestimate the role to be played by expanding the ecological knowledge of the population, particularly that of businessmen and economists. Strategic decisions and a daily commitment to conserve all resources and finished products while accelerating the manufacturing process represent our guarantee of future prosperity. Industry leaders, managers on all levels, and the public (great savings can be achieved in the domestic arena) must not only remember this but, I would say, feel it in their bones. Before building or expanding anything, it is important to ponder this question over and over: Cannot some savings (financial and environmental) be realized for the same or even less money? The final goal of manufacturing is, after all, not manufacturing for the sake of production, but rather the satisfaction of actual human needs. Our most urgent needs, here in the USSR and abroad, are to live under civilized conditions and to be confident of the wellbeing of our descendants.

THOMAS G. LAMBRIX

Global Climate Change: A Retrospective View from Business

Thomas G. Lambrix is director of government relations for Phillips Petroleum Company and chairman of the Global Climate Coalition, an association of companies and business trade organizations. Before joining Phillips over eight years ago, he worked for ten years with the federal government, holding positions at the U.S. Department of Interior and a presidential appointment to the White House domestic policy staff. Mr. Lambrix has an undergraduate degree in chemistry from Rutgers University and a masters in business from the University of Massachusetts.

I have heard it said that global climate change is the supertanker of environmental issues. It has all the characteristics of becoming the focus of the foremost environmental crusade of the 1990s, continuing well into the next century. In fact, some would argue it already *is* the principal environmental concern facing all nations of the world today, because the ecological implications from a future significantly or even moderately warmer than today are potentially substantial and would perhaps result in incredible changes to our physical world. These changes, should they occur, would also manifest themselves in economic, as well as societal, impacts that are hard to quantify, but would undoubtedly be dramatic.

Describing future global climate change is one of the most challenging undertakings our modern scientific community faces. Attempting to understand the world's climate involves grasping the innumerable interactive and complex variables that make up our global environment.

The political dimensions of global climate change are also extremely complex. Nations look at this matter differently—some with more concern than others, and some supporting responses that are more radical than others. As for the United States, over the past two and a half years its citizens have seen an array of stories in the media detailing a whole range of possible impacts associated with global climate change. They have heard from scientists, government representatives, and leaders of environmental groups. As a result, many Americans have become aware, and they are anxious about what global warming means for them and their children. While an informed public is desirable, in my opinion, most of the concern has been focused on the more pessimistic impact scenarios given visibility by the popular press.

Dr. Patrick J. Michaels, a climatologist at the University of Virginia, recently noted:

Time magazine painted the following picture in its 1988 "Man of the Year" issue, featuring planet Earth: If left unchecked, anthropogenerated alterations of the atmosphere will bring, by the years 2030–2050, a global temperature rise of 4°C, ecological chaos including famine, related civil strife, and tidal waves crashing through a Manhattan landscaped with palm trees. . . .

Such scenarios also abound in the writings of politically active environmental scientists, environmental lobbyists, and newswriters. United States elected officials have compared the current situation to that of Fascist Germany, where several events, such as Kristallnacht, presaged the holocaust. Those [who] do not recognize this have been labeled modern-day Neville Chamberlains. Draconian interventionist legislation has been proposed and the implementation of "global warming" is now a touchstone of U.S. foreign policy. Clearly, deep emotional commitments now guide this issue.[1]

There has been considerable attention to such views, but little public scrutiny has so far been devoted to what it might mean to adjust life-styles and otherwise pay for corrective measures in this country and around the world to address global climate change. So while the subject has been "teed up," awaiting action for some months, its full dimensions remain unclear and consensus on solutions is elusive. As I have said, global climate change is the supertanker environmental issue; it can potentially affect our values, our behavior, our social structures, and our institutions, which like a supertanker are slow to form, slow to stop, and slow to change. From the science to the politics, all of this, is incredibly complex. In commenting on the complexity recently, Dr. D. Allan Bromley, the president's science advisor, drew this humorous anecdote:

> In considering the challenge of understanding global change, it is easy to sympathize with the remark of the medieval king Alfonso X of Castile, who said, "If the Lord Almighty had consulted me before embarking on the Creation, I would have recommended something simpler."[2]

It is within the same context that U.S. industry wants to be an active and constructive participant in offering reasoned and responsible actions to address Alfonso's dilemma.

---

1. Patrick J. Michaels, "The Greenhouse Effect and Global Change: Review and Reappraisal," to be published in *International Journal of Environmental Studies*, 1990.

2. D. Allan Bromley, address to "Earth Observations and Global Change Decision Making: A National Partnership," conference sponsored by the National Aeronautics and Space Administration, the National Oceanic and Atmospheric Administration, and the Environmental Research Institute of Michigan, National Press Club, Washington, DC, September 18, 1989, pp. 9–10.

THE GLOBAL CLIMATE COALITION

Earlier in this century, President Calvin Coolidge announced, "The chief business of the American people is business." Today, promoting economic growth through private enterprise remains a chief business of this country, although there is a maturity to business actions that bears distinction. American companies today recognize the need not only to make a profit and enhance shareholder value but also to be responsible corporate citizens concerned about the quality of life in the United States and around the world. This has been true especially over the last twenty years, as businesses have invested in environmental quality, philanthropy, education, drug-abuse prevention, and other worthy social programs. I think today businesses realize they must be perceived accurately by the public as pursuing an agenda in harmony with society's goals. In particular, managers of U.S. industries appreciate that citizens no longer tolerate poor environmental quality, even in the name of absolute economic progress. Businesses spend many billions annually for environmental protection measures, and it is understood that more investments will be required. Past investments have already paid off, in demonstrable improvement to U.S. environmental quality.

In approaching the complex question of potential changes to the global climate, U.S. businesses take an active interest. The Global Climate Coalition, which sprang from that interest, is an ad hoc organization representing a broad spectrum of U.S. companies and trade organizations dealing in energy production and consumption, information, and allied industries. (A list of members appears on page 62.) Our involvement reflects the seriousness with which we view global climate change, and the potentially enormous impact certain suggested resolutions may have on our industrial base and the national economy. For the past year, the Coalition has embarked on programs to encourage scientific research, review voluntary initiatives that can be adopted, participate in governmental review of the issue both here and abroad, provide economic analysis, and otherwise promote an informed debate of policy responses.

The Need for Science

As far as supporting new government laws and regulations to deal with environmental matters, business has often been criticized for its opposition to new legislative proposals. At a minimum, business has been seen as a reluctant partner in seeking and achieving solutions—hiding behind a position of "more study." Critics see this stance as one of delay, of hope that the problems will just go away.

I am beginning to hear such criticism directed toward businesses for their position on global climate change, and I expect I will hear it again as the debate continues. This is unfortunate, because I believe it is only natural for industry to be interested in obtaining as many facts about a problem as it can before acting. This does not equal a lessening of concern but is a prudent posture. In everyday matters, business carefully pursues a study of data and options before making investments; this is the nature of business. And in the case of global climate change, despite the criticisms and the emotional appeal for drastic action, *more study is required.*

Most readers are familiar with the outstanding disagreements, questions, and uncertainties regarding the scientific basis for anthropogenic global climate change. While there is no disagreement over the greenhouse theory itself and the conclusion that atmospheric accumulations of greenhouse gases are increasing, there are differing views over the significance of future greenhouse gas emissions and whether and what should be done about gas emissions now. At a recent meeting of climatologists hosted by the National Academy of Sciences, for example, a group of nationally recognized climate experts generally agreed that emerging evidence has reduced expected global warming (responding to a carbon dioxide doubling) to about 2°C.[3] This is at the low end of the scale from the predictions publicized only a year ago. And *The Washington Post* recently cited the opinion of Mark Meier, of the Institute of Arctic and Alpine Research at the University of Colorado, that "the consensus now is that sea level will rise by about a foot in the next century, rather than the three feet that had been suggested in previous studies."[4] Even Dr. Stephen H. Schneider, the National Center for Atmospheric Research climatologist who has publicly stated a personal bias for moving ahead with policies to slow down the rapid buildup of greenhouse gases, wrote recently:

The consensus about the likelihood of future global change weakens over detailed assessments of the precise timing and geographic distribution of

3. National Academy of Sciences, panel on policy implications of greenhouse warming, Washington, DC, discussions of January 24, 1990.

4. William Booth, "New Models Chill Some Predictions of Severely Overheated Earth," *The Washington Post*, January 29, 1990.

potential effects and crumbles over the value question of whether present information is sufficient to generate a societal response stronger than more scientific research on the problems.[5]

It is therefore the opinion of the Global Climate Coalition that intensified scientific research should be our first priority. We should be focusing significant resources on reducing as much of the scientific uncertainty as possible over the next ten to twenty years. Obviously, more effort will be required but we believe this is a necessary and sound investment. Again, in Dr. Schneider's words:

A dedicated effort to accelerate the rate of progress could conceivably speed up the establishment of a consensus on regional issues, but at best ten years or so will be necessary even with a dramatic effort. *However, such efforts would clearly put future decision making on a firmer factual basis and help to make adaptation strategies more effective sooner.* [6] *(Emphasis added.)*

That is why the Global Climate Coalition supports the U.S. government's efforts undertaken by the Committee on Earth Sciences and believes it should move forward with all speed. The Coalition enthusiastically endorses President George Bush's call for substantially increasing federal funding for research devoted to all aspects of global climate change. In addition, the Coalition strongly believes government and industry should work together in accomplishing research goals. Every effort should be made to remove barriers to a greater working relationship between them and to identify specific opportunities for forming industry-government partnerships.

The Coalition recommends specifically that the United States focus its research on analytic efforts to establish the relative integrity of general circulation models; on understanding the dynamics of the carbon cycle; on sorting out the respective roles of clouds, oceans, and volcanoes; on developing reliable regional effects; and on determining the critical difference between real climate changes due to the normal variations of natural climatic phenomena and quantified changes attributable to anthropogenic gases.

While the Coalition firmly believes additional scientific understanding is necessary, it is important not to draw the conclusion that we are just hiding

5. Stephen H. Schneider, "The Greenhouse Effect: Science and Policy," *Science* 243 (February 10, 1989), p. 778.

6. Ibid., p. 781.

behind the "more study" solution. That is not an accurate representation. The Coalition vigorously pursues an expanded agenda, as described below.

Support for International Leadership

The Global Climate Coalition strongly supports leadership by the United States toward development of a responsible international approach to global climate change. Here we believe it is necessary for the U.S. government to exercise this leadership through its chairmanship of the Response Strategies Working Group of the Intergovernmental Panel on Climate Change (IPCC) and through the allocation of financial resources toward cooperative efforts here and abroad to address scientific, economic, and technological questions. The Coalition has met with various members of the IPCC's U.S. delegation, the Environmental Protection Agency, and the Departments of State, Energy, Commerce, and Interior. The Coalition also recognizes and supports the efforts of the many other countries, their industries, and nongovernment organizations involved in the IPCC process.

There is, I believe, little disagreement among interested parties over the premise that unilateral action by just one nation will not have a meaningful impact on the accumulation of greenhouse gases in our atmosphere. In fact, while North America and the industrialized nations of the world have been and are today the major contributors to anthropogenic greenhouse gas buildup, this trend is declining and will reverse itself in the future, when the now developing world becomes the significant contributor. Cooperation and active input from developing countries will be essential for an effective global response to potential global warming.

In a report that the Global Climate Coalition commissioned in 1989 to evaluate some of the economic implications of unilateral action on global warming, the Washington, D.C.–based Center for Strategic and International Studies reaffirmed the importance of this conclusion:

The world's population is expected to grow from 5 billion in 1988 to 6 billion in the year 2000 and 8 billion by 2025. Most of this increase will be in the Third World and the Communist nations, where energy demand has doubled since 1971. Almost all of this increase has been in fossil fuels.

Developing countries are expected to quadruple their energy consumption between now and the year 2025. By the middle of the next century *they will account for the bulk of the greenhouse gases emitted into the atmosphere, even if they succeed in doubling their energy efficiency.* [7]

The challenge to companies and policymakers in the United States and abroad is to work with developing nations as they strive to improve the standard of living for their expanding populations and to do so in a way that promotes economic growth and sustains the environment—including the global atmosphere. I am confident U.S. businesses will be anxious to help, particularly in the search for equitable means to enhance the transfer of enery-efficient technologies.

Support for Initiatives

At home, the Global Climate Coalition advocates strong leadership by the president and Congress regarding global climate change. We support particularly President Bush's endorsement of a free-market approach in identifying appropriate responses. We also agree with President Bush that unilateral intervention by the United States in the national, or global, economy is not warranted at this time. This does not mean other activities cannot be pursued. In addition to backing the president's announced program for enhanced scientific research, for example, the Coalition enthusiastically supports investments in cost-effective energy efficiency, orderly phaseout of chlorofluorocarbons, development of public education programs, recycling, wast minimization programs, and implementation of reforestation initiatives.

We believe there are yet other opportunities to demonstrate presidential and congressional leadership. I will mention just two specific programs. First, the federal government should lead by example and invest in energy efficiency and renewable energy technologies in the federal estate. This will educate the public at large about the application of these technologies and help reduce production costs. Second, the president should announce a nationwide recognition program designed to identify and reward citizen efforts and investments in such areas as new energy-efficient technologies, school programs (here the rewards might take the form of scholarships for science and related subjects), and community efforts at recycling and, for example, reforestation. A town in the United States might "adopt" a town in another country and set up information and assistance exchanges to emphasize local actions by citizens around the globe. The president could

7. Center for Strategic and International Studies, "Implications of Global Climate Policies," report for the Global Climate Coalition, Washington, DC, June 27, 1989.

instruct the chairman of the Council on Environmental Quality, in cooperation with other appropriate agencies, to develop and implement such a network. A congressional joint resolution regarding these local efforts would also be welcome. Companies could set up parallel programs; in fact, many companies are already taking specific action on their own.

In an address to a recent meeting of the National Governors Association, Secretary of State James Baker discussed the Bush administration's program to address global climate change. He described it as a "no-regrets policy," and explained the administration was prepared to take actions that are "fully justified in their own right *and* . . . have the added advantage of coping with greenhouse gases."[8] The Global Climate Coalition supports this approach to policymaking on global climate change and recognizes that business has a responsibility for developing constructive responses. In order to establish itself as a credible, informed, and committed partner, business must increase its efforts to make environmental conscientiousness a matter of routine practice and continue to accelerate its own efforts to invest in cost-effective energy-efficient technologies. Coalition member companies have accrued almost twenty years of research and technological experience in the pursuit of energy sources that serve sound economic and environmental policies equally. We believe we are making progress, and we intend to continue looking for ways to increase efficiency and reduce pollution. We hope that in the future any improvements we develop in this country can be made available around the world.

The Global Climate Coalition is actively investigating a menu of industry initiatives that can be supported now: these make sense in their own right by providing cost-effective investments and do not aggravate the buildup of greenhouse gases. Voluntary industry adoption of specific programs could include energy-efficient technologies, productivity increases, development of new products and technologies, waste minimization, and identification of innovative programs to disseminate technology and deal with the real problems of technology transfer from the developed to the developing world. The point to stress here is that it is critical for the U.S. government to create a favorable setting to encourage business. And for the private sector to be most effective in responding, government policy should rely heavily on free-market principles and the application of incentives *rather* than the institution of penalties. We need to stress flexibility in our approach to energy, agriculture, foreign assistance, and research so that we can adjust our programs and investments as an understanding of global climate change increases, and as our multilateral discussions mature.

8. James A. Baker III, "Diplomacy for the Environment," presentation to the National Governors Association, February 26, 1990, pp. 5–6.

We are pleased to see President Bush support this flexibility as well. As he stated to the Intergovernmental Panel on Climate Change:

> If we hope to promote environmental protection and economic growth around the world, it will be important . . . to work [not in conflict with but together] with our industrial sectors. That will mean moving beyond the practice of command, control, and compliance, toward a new kind of environmental cooperation.[9]

CONCLUSION

Global climate change has achieved high visibility on the environmental and political agenda of all industrialized nations of the world and many developing ones. In essence, this has come to pass quite rapidly, over a period of less than three years. The result: an incredible upswell of political momentum to call for "solutions," at a rate that has outpaced our understanding of the scientific causes of the phenomenon and the economic impacts of our possible responses. While scientific attention to the concern is widespread and will continue, science still lags behind the pleas of politicians for its assistance in drafting a reply. Even though more recent scientific forecasts have lowered rather than raised the range of potential warming and ecological damage, pressures continue in the U.S. and abroad for governments to do something now.

After two years of intense public and political attention to the projected ecological disasters caused by global climate change, we, as a global community, are only now beginning to address ourselves to the economic consequences of responses some have touted as absolutely necessary to deal with this "crisis." The question of who pays and who benefits is already becoming the principal negotiating point as the United States and nations around the world determine economic and social effects of mitigation and adaptation responses. I don't believe that Americans will continue to view global climate change in the same context as other environmental topics, such as enactment of new clean-air legislation. Global climate change is too complex and the requirements on our citizens much more drastically felt if severe intervention strategies are implemented. This point was specifically noted in a recent report by the Department of Energy, which concluded:

> Unless . . . there is uncontentious sustained evidence of climate-change impacts, a real possibility exists that the current appearance of societal consensus in favor of urgent action on climate change may fragment as

9. George Bush, remarks to the Intergovernmental Panel on Climate Change, Washington, DC, February 5, 1990, p. 3.

the costs to various sectors of society become clear. Even if the resulting economic disruption and societal conflict can be overcome, such conflict carries its own costs which must be considered part of the true social cost of a policy package.[10]

To address important economic issues, the Global Climate Coalition has retained a team to undertake an economic assessment of potential near-term policy responses by the United States. Specifically, the objectives of the study are to

- analyze the impact of proposed policy actions on the U.S. economy and on the competitiveness of U.S. industry in world markets;
- determine the impact of proposed policy actions on net global greenhouse gas emissions;
- assess the geopolitical context of proposed policy actions by the United States.[11]

The Coalition's hope is that the results of this study, when published, will advance an understanding of economic ramifications to be considered by our policymakers.

Many sectors of society—academic think tanks, environmental groups, climatologists, businesses, government bureaucrats, elected officials, the media—generate information themselves, and are being overwhelmed by others with a mass of materials advocating a full spectrum of views. Sorting out the tidbits of valuable information among the noise is a real challenge. All of us need to be alert to other agenda and political motivations that will struggle and compete with each other to influence and drive the issue.

Despite all of the hype, however, the notion of potential climate change brought about by global warming remains a serious consideration. I am optimistic we will see responsible and balanced proposals crafted to understand and respond. As I have said, to be effective and credible, these proposals must be developed multinationally and must be based on sound science.

Industries are moving to ensure their pursuits and operating principles are compatible with the goals of society, including those for preserving and enhancing the quality of the environment. This trend will continue, in the United States and internationally. Business *does not* oppose action. But we in business do object to acting in haste when the consequences can be

10. U.S. Department of Energy, "A Compendium of Options for Government Policy to Encourage Private Sector Responses to Potential Climate Change," report to the Congress of the United States, October 1989, vol. 2, pp. 12–14.

11. The Global Climate Coalition has retained the team of the Center for Strategic and International Studies; Putnam, Hayes & Bartlett; the International Resources Group; The Brock Group; and DRI/ McGraw-Hill to undertake an economic assessment of potential near-term policy responses by the United States in regard to reducing emissions of greenhouse gases.

enormous and the intended result ineffective. By necessity, business will look at responses to global climate change from the perspective of economic reality. We want our actions to be supported by science, and we need to understand the tradeoffs between benefits and costs. As President Bush indicated to the delegates to the IPCC:

> Much remains to be done. Many questions remain to be answered. Together, we have a responsibility to ourselves and the generations to come to fulfill our stewardship obligations. But that responsibility demands that we do it right.[12]

Rapid changes are occurring in our global economic environment. Now more than ever we live in a global economy. International competitiveness will be a preeminent factor in determining not only corporate success but a higher quality of life for our citizens. How our response strategies to global climate change affect our competitiveness must be a major consideration. The Global Climate Coalition's view is that strong economies will be necessary to provide the capital for investment in needed technology and for taxpayer-financed assistance to other nations. The past year or so has seen the demise of the centrally planned economic systems of Eastern Europe, as citizens of these nations cried out for freedom and a better life. They are now endorsing free-market economies. Our government policies should stimulate investments by the private sector in these countries and in the developing world to help improve both their economies and their environments.

POSTSCRIPT

The Greenhouse/Glasnost conference held at Sundance, Utah, in August 1989 brought together Soviet and U.S. scientists, politicians, academicians, environmentalists, industry leaders, and others, to share concerns about the implications of possible climate changes brought about by global warming. I was glad to have had the opportunity to participate there—to listen and relate the views of business on this very important matter. I think I learned, and I hope I was able to convey business's ideas and recommendations. Of the many issues discussed, one stands out in my mind. It was a notion referred to in a speech by the Cornell University astronomer Dr. Carl Sagan.[13] He was giving an overview of the Voyager program as the satellite

12. Bush, p. 3.
13. Dr. Carl Sagan, remarks to the Greenhouse/Glasnost Conference, Sundance Symposium on Global Climate Change, Institute for Resource Management/Soviet Academy of Sciences, Sundance, Utah, August 1989.

was nearing its approach to the planet Neptune. I recall Dr. Sagan's praising the accomplishments of this project, indicating that Voyager had traveled to the outer limits of our solar system, achieved far more than its objectives, and was completing its mission on budget. Dr. Sagan indicated it was wonderful what technology could do when it was freed up to meet such a formidable challenge.

Given the same opportunity to unleash the creativity of our enterprises within free and strong economies, the Global Climate Coalition believes that the United States and other nations of the world will likewise respond to the scientific and political challenges posed by global climate change.

GLOBAL CLIMATE COALITION
MEMBERSHIP

Aluminum Association
American Electric Power Service
   Corporation
American Gas Association
American Iron & Steel Institute
American Mining Congress
American Nuclear Energy
   Council
American Paper Institute
American Petroleum Institute
Amoco Corporation
ARCO
Armco, Inc.
Association of Home Appliance
   Manufacturers
AT&T
Automobile Importers of America
BHP–Utah Minerals
   International, Inc.
Business Roundtable
Champion International
Chemical Manufacturers
   Association
Chevron U.S.A. Inc.
Chrysler Corporation

Coalition Opposed to Energy
   Taxes
Consolidation Coal Company
Consumers Power Company
Council of Industrial Boiler
   Owners
Dow Chemical Company
E. I. Du Pont de Nemours &
   Company
Eastman Kodak
Edison Electric Institute
Electricity Consumers Resource
   Council
Entergy Corporation
Ford Motor Company
Fusion Power Associates
General Motors Corporation
Georgia-Pacific Corporation
Hercules Inc.
IBM
Illinois Power Company
International
   Business-Government
   Counselors, Inc.
Jefferson Energy Foundation

Kaiser Aluminum & Chemical
    Corporation
Maytag Corporation
Monsanto Company
Motor Vehicle Manufacturers
    Association
National Association of
    Manufacturers
National Coal Association
National Steel Corporation
Occidental Chemical Corporation
Pacific Gas & Electric Company
Peabody Holding Company, Inc.
Petrochemical Energy Group
Petroleum Marketers Association
    of America
Phillips Petroleum Company
Portland Cement Association

PPG Industries
Process Gas Consumers Group
Public Service Indiana
Rubber Manufacturers
    Association
Shell Oil Company
Society of the Plastics Industry,
    Inc.
Southern Company Services, Inc.
Texaco, Inc.
Union Carbide Corporation
United States Chamber of
    Commerce
United States Council for Energy
    Awareness
United States Council for
    International Business
UNOCAL Corporation

Changing Climates, Changing Habits: Energy Efficiency and the Global Warming Prevention Act

Claudine Schneider is a five-term congresswoman representing Rhode Island's second district. She has written, sponsored, and helped to pass a wide range of key environmental legislation, including bills to end ocean dumping of sewage sludge, to direct the president to negotiate an international treaty to conserve biological diversity, to fund a national radon-testing program, and to create economic incentives for energy efficiency and waste reduction in manufacturing. Representative Schneider is the author of the widely acclaimed Global Warming Prevention Act (HR 1078), which she introduced at the end of the 100th Congress. This act calls for the employment of market incentives to encourage industry to reduce greenhouse gases, to use renewable resources, and to implement general policies of energy efficiency and energy-efficient technologies. Representative Schneider received an award from the Center for Environmental Education for her work on the Endangered Species Act of 1987, and the Conservation Advocate of the Year Award in 1989 by the National Wildlife Federation.

The year 1988 has the distinction of being the year global warming captured the public's attention. Americans witnessed an unusually large number of disasters, all symptomatic of what global warming portends. These included a crop-destroying drought that devastated many American farms; raging forest fires that spread uncontrollably through our commercial timber and pristine wilderness areas; Hurricanes Gilbert and Helene, which ravaged the Caribbean and our southeastern shores and left entire communities homeless; a record heat wave that gripped the nation and intensified the unhealthy effects of urban pollution; and the tragic flooding of Bangladesh, which left a legacy of disease and starvation.

Scientists lack adequate data to say whether these costly incidents were actually influenced by a global warming trend. But the scientific community is in large agreement in warning us that such disasters will occur with increasing frequency and severity in the future, if greenhouse gas emissions rise unchecked. According to the National Academy of Sciences, maintaining "business as usual" will lead to a temperature rise of 4° to 9°F in coming decades. Clearly, more basic research on climate change is paramount if science is to predict better the many consequences from rising levels of greenhouse gases.

At the same time, the risk of such dire consequences have led several of the world's most prestigious scientific bodies, including the World Meteorological Organization and the International Council of Scientific Unions, to call for steps to prevent a worsening of the global warming trend. Minimum reductions of 20 percent from current carbon dioxide (CO_2) emission levels within the next fifteen years was the recommendation issued at the 1988 Toronto World Conference on the Changing Atmosphere.

This call for action, however, is challenged by a number of highly vocal cynics and skeptics. The thrust of their argument is this: The vast increase in human-generated greenhouse gases might be absorbed by the ocean and other ecosystems, and thus there might be very little global warming. Therefore, any action is opposed on the grounds it would incur needless, burdensome expense.

How are policy decisions to be made in the wake of this seeming uncertainty and controversy? An obvious way to deal with rising levels of greenhouse gases is that in which we enact policies to deal with toxic chemicals or the threat of cancer. Great uncertainties characterize both these matters, but despite uncertainty society has established certain standards that limit human exposure to harmful levels of toxic chemicals, even to the extent of outright bans on some of the most pernicious ones. Likewise, vigorous government research efforts on preventative measures continue, and information is generally disseminated to the public on how to avoid suspected carcinogens.

In a similar manner, society can phase out greenhouse gases with nonpolluting or less polluting methods that provide the same economic services. The phaseout of ozone-depleting chemicals, which are also greenhouse gases, is a case in point. Substitute refrigerants, cleansing agents, foam materials, and so on are being used instead of or developed to displace chlorofluorocarbons and other damaging halocarbons. The Montreal Protocol, signed by the United States and other nations in 1987, already requires a 50-percent phaseout over the next decade, and Congress is close to enacting a complete phaseout.

Without a doubt, the energy sector offers the biggest opportunity for displacing the bulk of greenhouse gases. The combustion of fossil fuels contributes about half of all greenhouse gases, most notably carbon dioxide. Over the past several years a series of congressional hearings has focused on the various options for a shift away from the most polluting fossil fuels, oil and coal. It is quite apparent that while the options are many, they range in cost by tenfold or more. It is therefore imperative that governments formulate global warming prevention policies and strategies that do not entail needless and wasteful expenditures by taxpayers, rate-payers, and stockholders.

Legislation I have authored advocates just such a strategy. The Global Warming Prevention Act (HR 1078), currently cosponsored by more than a third of the House of Representatives and endorsed by forty national organizations, lays out a course of action even cynics and skeptics should find acceptable. It is based on the timeless adage popularized by one of our country's founding statesmen, Benjamin Franklin: "An ounce of prevention is worth a pound of cure." The guiding principle of this bill is to implement a National Least-Cost Energy Policy, which would rank our energy options in order of their cost-effectiveness and environmental impact.

The bottom line: to engender a positive-sum strategy that cuts greenhouse gas emissions while cutting energy costs too. Indeed, if one judges from the available evidence of what least-cost energy offers, the bill could easily have been called the "Global Competitiveness and U.S. Productivity Enhancement Act." The several dozen policy measures detailed in the legislation should eventually help save several hundred billion dollars per year on consumer energy bills; create high-efficiency energy-generating and energy-consuming products and services for export; reduce foreign oil imports and the trade deficit; and reduce a range of other environmental pollutants in addition to greenhouse gases.

The message is repeated in testimony delivered by numerous expert witnesses before Congress who have detailed the preeminent role energy efficiency has come to play in our economy. Improvements in energy efficiency, which flow from scientific advancements and technological innovations that enable more energy to be delivered with less energy input, have become the key source of low-cost energy services to industry and consumers.

It is a too little appreciated fact that over the past fifteen years investments in improving the efficiency of America's buildings, appliances, vehicles, industrial equipment, and other energy-consuming devices have cut energy consumption by one-third, reduced carbon dioxide, and sulphur and nitrogen oxide pollutants to 50 percent below what they would have been, and trimmed the U.S. energy bill by a phenomenal $160 billion per year. Success breeds success, and this is undeniably true about remaining opportunities to save even more money and cut greenhouse emissions through additional efficiency improvements.

Detailed government and private studies show the U.S. economy could maintain its robust economic growth while achieving $200 billion per year in energy savings through continued investment in efficiency technologies. These investment opportunities are on the average two to five times cheaper than conventional fossil and nuclear resource options. Declining levels of greenhouse gas emissions become a derivative, cost-free benefit, and efficiency improvements would greatly enhance our economy's productivity and competitive edge in the global marketplace. The United States, it should be remembered, currently requires twice the energy to produce a dollar of Gross National Product than foreign competitors such as Japan and many West European nations require.

Let me illustrate several opportunities for energy efficiency. Consider lighting, and the associated air-conditioning needed to remove the heat generated by lights, which consumes the equivalent of nearly half of all coal burned by utilities. A rich variety of highly efficient lighting products— compact fluorescent lamps; dimmable solid-state electronic ballasts, daylighting sensors, occupancy sensors, imaging specular (mirrorlike) reflectors, polarizing lens, and so on—now make it possible to deliver the same lighting amenities while consuming half to three-fourths less electricity. These more efficient lighting products would annually cut carbon emissions by tens to hundreds of millions of tons and reduce energy bills by several tens of billions of dollars.

The compact fluorescent lamp is a technological marvel. It provides the same light as an incandescent bulb, which requires four times more electricity, and it lasts ten times longer. Over its lifetime, an 18-watt compact, displacing a 75-watt incandescent, will save the consumer between $30 and $50, and prevent the combustion of a ton of CO_2. A compact-lamp factory is even more impressive. A small-sized factory producing 2.5 million lamps per year costs $7.5 million to set up, but then displaces the need for a 350-megawatt plant costing over $350 million to construct! Two fair-sized factories, with a combined output of 10 million lamps per year, would constitute a top-fifty coal mine in the United States, the key difference being that the factories would prevent the combustion of 3 million tons of coal and save customers $500 million.

Electric motors and industrial drive devices consume 60 percent of U.S. electricity; the electrical costs worldwide exceed $300 billion per year. A range of high-efficiency motors and drive equipment now offer to deliver the same industrial drive services while saving between one-third and two-thirds of the power input. According to the Rocky Mountain Institute's Competitek Industrial Drive analysis, full use of properly sized high-efficiency motors, adjustable speed drives, fast-speed controllers for turbomachinery, power factor controllers, and similar devices would cut the U.S. electricity bill by $25 to $50 billion per year, and eliminate the need for between 80 and 190 giant power plants. The bottom line for business is an increased profit sheet through lowered operating expenses, while the gains for society are tax-free and cost-free reductions in greenhouse gases.

The refrigerator offers another outstanding opportunity to cut costs *and* pollution. The average refrigerator combusts its volume in coal each year, while a highly efficient one reduces that volume to a vegetable bin's worth. Food cooling consumes nearly 25 percent of U.S. residential electricity, the output of about 25,000 megawatts. Whirlpool, Amana, General Electric, and other manufacturers currently produce refrigerators requiring half the electrical consumption of comparable models, with the extra cost-efficiency improvements paid back to consumers within several years through reduced utility bills. Other models, such as the Sunfrost refrigerator marketed out of California, reduce electric consumption by up to 90 percent. The Sunfrost is currently hand-manufactured and sold only to rural households installing photovoltaic (solar-cell) power systems. Despite its hefty price, triple that of conventional models, it saves rural households $10,000 to $20,000 on photovoltaic arrays. When mass-manufactured, the Sunfrost is expected to cost no more than inefficient models.

Despite the enormous potential for cutting gas and electric bills by half to three-fourths, a number of institutional barriers and market imperfections inhibit the timely use of the products described above. Federal energy subsidies and state utility regulations encourage the construction of the power plants over better cost-efficiency options by industry. Subsidies should be eliminated or shifted to create a level playing field. And innovative regulatory policies should reward utilities for helping their customers save money by saving energy, whenever this is less costly than building new power plants.

Fortunately, the National Association of Regulatory Utility Commissioners strongly supports these policies, known as Least-Cost Utility Planning. LCUP was pioneered in the northwestern United States after Congress mandated the process in 1980. One of the cutting-edge, and most exciting, applications of LCUP is the "collaborative process" now implemented by former adversaries, such as various New England electric utilities on the one hand, and the Conservation Law Foundation on the other. The project bodes well as a model for other parts of the nation to emulate. Energy-saving programs have become the utilities' top priority for satisfying growing energy demands.

The transportation sector holds enormous potential for energy savings. The nation's more than 130 million light cars and trucks now average less than 20 miles per gallon. These vehicles consume over 7 million barrels of oil per day, emit their weight per year in carbon emissions, and are responsible for about one-third of U.S. greenhouse gas emissions. Federal fuel economy standards have raised new vehicle fleet averages to roughly 27 miles per gallon, nearly twice the level at which they were enacted in the mid–1970s. Those standards have cut carbon emissions by millions of tons per year, while accumulating net energy savings for consumers of $180 billion as of 1987.

Unfortunately, since less than one-fifth of total automotive costs is fuel, consumers have little incentive to purchase vehicles with much more efficiency than 30 or 35 miles per gallon. This remains the case even at fuel prices two to four times the U.S. pump price, as are found in most European nations. Yet at least seven manufacturers have already tested cost-effective, high-performance prototype cars that get 65 to 135 miles per gallon.

Obviously, from a global societal perspective, mass production of these highly efficient vehicles would greatly reduce greenhouse gases (as well as urban smog, acid-rain pollutants, vulnerable foreign oil imports, and our $130 billion trade deficit). Policy actions are available that overcome the market's failure. The Global Warming Prevention Act includes three: a 65-percent increase in the federal vehicle energy efficiency performance standards over the next decade; tax rebates of up to $2,000 per car as an incentive for consumers to purchase more efficient vehicles; and an increase in the current gas-guzzler tax on very inefficient vehicles. These three measures are certain to raise U.S. efficiency averages to 45 and 35 miles per gallon for light vehicles and light trucks, respectively, by 1999. The result prevents the combustion of 15 billion barrels of oil over the next thirty years.

These automotive improvements will save consumers tens of billions of dollars at the gas pump, prevent the release of over a billion tons of carbon into the atmosphere, improve urban air quality by reducing volatile organic compounds, and continue to save an additional 500 million barrels of oil each year. It is also noteworthy that these modest improvements exceed the combined oil resources extractable from the Pacific Coast, Alaska, the Arctic National Wildlife Refuge, and the Atlantic Coast. And this by no means exhausts all economically attractive efficiency improvements: an additional 7 billion barrels could be saved by achieving light car and light truck levels of 60 and 45 miles per gallon, respectively, early in the next century.

Energy efficiency is not a panacea to the global warming problem, but it does represent the most important initial step this and other nations can take in reducing greenhouse gases in a cost-minimizing manner, while spurring economic productivity at the same time.

These energy savings may offer one of the few new sources of investment capital for undertaking additional measures to control greenhouse gas emissions, especially for capital-starved developing countries. The global energy efficiency study conducted by a highly respected international research team, published as *Energy for a Sustainable World* (Wiley, 1988), determined that it is cost-effective to maintain a 3-percent annual rate of energy efficiency improvement for several decades to come. Worldwide energy savings would eventually exceed a staggering $5 trillion per decade.

In the absence of efficiency gains, *all* of the following energy supply expansion would have to occur over a fifty-year period: construction of a 1,000-megawatt coal plant every two days and a 1,000-megawatt nuclear plant every four days; installation of the equivalent of an Alaska pipeline every other month; increase in OPEC output to maximum capacity.

Not only would efficiency gains obviate the need for most of this expansion, accruing enormous capital savings, but the energy conserved would prevent some very serious environmental and security problems. Carbon emissions would decline from today's 5 billion tons per year, instead of tripling to 15 billion tons per year. The use of weapons-grade plutonium in the nuclear fuel cycle would be unnecessary, as would the annual shipment to and from reactors of 5 to 10 million pounds, or 500,000 atomic bombs' worth (50,000 bombs are currently in the world's nuclear arsenals). Oil prices would remain weak, eliminating the sudden price surges historically experienced whenever OPEC has exceeded 80-percent capacity.

Milking this cash cow is imperative, when we consider the equally staggering debt and deficits the United States and many other countries face. Without these relatively easy savings, it may be impossible to finance the billions of dollars of additional research and initiatives that must be undertaken. These include immediate prevention of deforestation; promotion of reforestation, agroforestry, and high-yielding, low-input agriculture practices worldwide; eliminating stratospheric ozone-depleting chlorofluorocarbons (CFCs) and halocarbons, and reducing other greenhouse gases (e.g., methane and oxides of nitrogen); achieving population stabilization, sooner rather than later, through promotion of universal family-planning services; and rapid development of ecologically sustainable solar-energy and renewable-energy resources.

Halting deforestation must be a top priority of the global community. Twenty-five million acres of forests—about the size of Great Britain—are destroyed each year, and their destruction contributes 10 to 20 percent of the greenhouse gases loaded into the atmosphere annually. Reforestation occurs at a miserably slow rate—only a few percent of the trees cut in Africa are replanted, 10 percent in Latin America, and about 25 percent throughout Asia and the United States. Reforestation must be greatly increased, but preservation of existing forests is even more of an imperative. As a paper by Dr. Dan Botkin of the University of California at Santa Barbara and Dr. Art Rosenfeld of the Lawrence Berkeley National Laboratory recently explained, destroying an acre of forests releases 52 tons of carbon, while a newly planted acre only sequesters .52 tons of carbon per year. "In summary," they write, "it takes 100 years of growth to offset the damage of clearing the same area of forest."

One intriguing proposal to preserve threatened forests would require utilities to offset carbon emissions from power plants by making contractual agreements with governments to protect designated forest areas from being cut down. Such "debt-for-nature" swaps already exist in some countries.

Estimates at Lawrence Berkeley Laboratory indicate that offsetting fossil emissions from a power plant would probably amount to less than 1 cent per kilowatt-hour generated. The average cost per kilowatt-hour in the United States in 1988 was 7 cents.

Boosting the efficiency of wood-consuming devices, whether they are the stone fires still used by over 1 billion people worldwide, or those used by agroindustries that burn biomass wastes, can greatly diminish pressure on forests. The authors of *Energy for a Sustainable World* estimate that replacing inefficient stone fires with efficient stoves that require half the wood or less would conserve enough wood to generate 160,000 megawatts of electricity, while saving Third World families the billions of hours spent each year collecting it.

For agroindustries, the United States has produced a superior generating technology, the advanced gas turbine. The turbine was originally researched and developed over the past decade as an engine for Air Force aircraft, but it has been modified for use as a highly efficient electric and steam cogeneration unit. The more advanced gas turbines operate 50 percent more efficiently than conventional power plants, and they cost half as much to build as new coal or nuclear plants.

A study by the U.S. Agency for International Development found that this aircraft-derived gas turbine could meet the steam-processing needs of sugar cane factories in seventy developing countries cost-effectively, while also cogenerating an estimated 50,000 megawatts of excess electricity. This is equal to one-fourth of the electricity currently generated in these countries. Most important, the turbines can use sugar-cane wastes, renewably cultivated tree crops, and other biomass as fuel, instead of relying on the next "least-cost" fuels, imported oil or coal. Thus fuel-wood plantations help prevent global warming, as trees serve a multiple function in restoring carbon to the soil, sequestering carbon in the trees' biomass, and displacing a fossil fuel.

Recent congressional testimony indicates that aeroderived turbines offer an excellent opportunity for U.S. farmers to diversify their income by planting tree crops. The 120 million acres of U.S. lands classified as eroded or eroding are capable of growing biomass fuels sufficient to replace over two-thirds of current coal-fired power plants at the time of their retirement.

At another level, pioneering research by Lawrence Berkeley Laboratory and the American Forestry Association (AFA) have found community and urban reforestation to be among the lowest-cost options available for offsetting carbon emissions. American cities have been deforested as a result of urban expansion over the past half-century, and currently only one tree is replanted in the United States for every four that die or are cut down. As communities are less shaded, more fossil fuels are burned to operate air conditioners. Lawrence Berkeley Laboratory estimates that 20 percent of smog incidences are related to the "urban heat island" caused by fewer shade trees.

The AFA's Global Releaf Program shows American communities and citizens how to go about planting between 160 and 300 million trees, which it has reported could be placed selectively around buildings, on streets, and in community parks. The trees would help communities save over $1 billion per year in air-conditioning costs and thus help reduce carbon emissions by tens of millions of tons per year. A single tree planted in an urban setting has the same offsetting effect as fourteen planted in rural areas, because of the reduced need for air-conditioning.

In his 1990 State of the Union address, President George Bush stressed the importance of tree planting. The president called for the planting of 1 billion trees per year in communities and rural areas across the nation, and he included roughly $200 million in his budget to initiate this effort.

We *must* give high priority to reinvigorating the nation's energy efficiency programs and renewable-energy programs; their budgets were reduced 50 percent and 75 percent, respectively, over the last ten years. Research into and development of these non-greenhouse energy options receive a mere $160 million and $110 million, respectively, of the U.S. Department of Energy's $16.5 billion yearly budget.

These low funding levels are unfathomable, especially considering that the federal energy efficiency R&D program has been a spectacular success. A review by the American Council for an Energy Efficient Economy concluded that if just seven of the most successful projects had to justify every tax dollar ever expended on energy efficiency R&D, taxpayers would still realize a $50 return on each and every dollar. Yet according to a mid–1980s update by the American Institute of Physics, which pioneered assessments on energy efficiency potentials in the U.S. economy, the nation has just scratched the surface.

In the case of renewables—solar, biomass, hydro, wind, and geothermal resources—the United States has barely tapped its massive reserves, although they do account today for 10 percent of the nation's total energy sources (50 percent more than does nuclear power, which receives five times more federal R&D funds than nonpolluting, nonhazardous renewable resources).

According to the Department of Energy's Energy Research Advisory Board (ERAB), the nation's reserves of renewable energy resources exceed

80,000 quads (a quad is a unit of energy equal to 1 quadrillion British thermal units, or BTUs). This is a thousand times the total U.S. energy consumption in 1989, and five to ten times larger than U.S. coal reserves. The ERAB estimates that with a modest but stably funded R&D program, our nation could economically extract over 80 quads from this new resource base, that is, 75 percent of the amount of energy it is projected we will need within the next 25 years. We should take this insight seriously.

Much of the nation's garbage wastes represent an untapped renewable resource. We are long overdue in alleviating a nagging problem of waste disposal by spurring greater reliance on reusable and recyclable materials. Studies have shown that if the United States increased its recycling rate 10 percent above the projected level by 1992, and 30 percent above the projected level by 2008, the nation could accumulate savings of $30 billion and 7 quads of energy.

This strategy is well suited for reducing America's 1 billion tons of annually generated hazardous and toxic pollutants. According to reports prepared for Congress by the Office of Technology Assessment, up to 50 percent of all hazardous wastes could be eliminated with existing technology in the next few years. In addition to multiple environmental benefits, U.S. industry and taxpayers would save tens of billions of dollars each year in manufacturing, regulatory, and liability costs.

I would like to recall the findings of a 1981 federal report, *Energy and Air Quality,* that a significant increase in atmospheric carbon dioxide "is expected to be caused by world-wide increases in population which will increase energy use and speed deforestation, limiting the control of CO_2 by natural processes." As World Bank president Barber Conable recently reiterated, unchecked population growth threatens to overwhelm even the best of development assistance programs. The World Bank and other international organizations have repeatedly stated the imperative of stabilizing global population through universal family-planning services. Achieving this goal in a reasonable time frame could affect global warming as strongly as any other kind of action.

Worldwide expenditures for family-planning assistance amount to $3.2 billion per year, yet 55 percent of women in the world presently do not have access to some form of birth control; this includes 38 percent of U.S. adult women in need of it, who cannot afford planning services. According to a World Fertility Survey by the United Nations, universal access to voluntary family-planning services could result in a one-third decline in Latin American and Asian births and a one-fourth drop in African births. Worldwide population levels would stabilize short of 10 billion, instead of surging to a projected 14 billion over the next century. To achieve this lower goal will require raising the funding level to $10.5 billion by the year 2000. The current U.S. commitment of $230 million would have to increase to $1.1 billion by that time, roughly the cost of a six-pack of beer per person in the nation. The United States cannot find a better use of its assistance dollars, not only for preventing environmental problems but also for alleviating poverty and spurring economic development. United Nations studies show that each dollar expended on these services reaps a sevenfold return on the taxpayers' investment in the form of improved maternal and infant health and improved productivity.

We must act expeditiously and comprehensively to slow greenhouse warming, and we can do so in a cost-minimizing manner that could help alleviate a whole variety of pressing national concerns and global problems in the process. It will take various changes in public policies at every level of government, local to international. But preventing a worsening of the world's climate will depend on the commitment of individuals to take personal responsibility. Global change will result from the accumulation of such simple and straightforward actions as replacing incandescent light bulbs with compact fluorescent lamps, driving more efficient cars and trucks, and recycling bottles and paper. Paper, for example, consumes half of all harvested wood worldwide. Yet less than 25 percent of it is currently recycled from consumer waste.

Perhaps never before in history have individual actions come to play so great an influence on our physical world and its future. Prevention pays, and it is incumbent on each and every one of us to capitalize on these opportunities. Such good stewardship will prevent the squandering of our precious natural endowment.

Citizens have daily opportunities to honor and heal the environment, in the way they purchase, consume, vote, and invest. Citizen empowerment can be exercised: demand that local utilities offer customers rebates to purchase energy-efficient devices, that the stores you patronize support labels on products to prove their environmental acceptability, and that elected officials actively pass legislation like the Global Warming Prevention Act.

HOWARD P. ALLEN

What Are the Industrial Considerations in a Climate of Global Warming?

Howard P. Allen is the chairman and chief executive officer of SCEcorp and its electric utility subsidiary, Southern California Edison Company. His many directorships include the Association of Edison Illuminating Companies, the Edison Electric Institute, and the Pacific Coast Electrical Association.

Mr. Allen has been the recipient of many awards and citations for public service from the State of California and the County and City of Los Angeles, and the 1988 Brotherhood Award from the National Conference of Christians and Jews. He serves as a trustee and member of the executive committee of the Los Angeles County Museum of Art.

Mr. Allen joined Southern California Edison in 1954. He is director and chairman of the board of the Institute for Resource Management.

Whenever thoughtful men and women gather to discuss the future, there will be disagreements on many fronts. But I think we can start with one premise: We all want to leave our children and our children's children with a life that is fuller, richer, and more bountiful than ours. And that includes an environmental quality equal to, if not better than, that which our parents and our parents' parents enjoyed.

Our grandparents did not think of the environment in the same way we do today. These are different times. Today concerned people worry whether the Earth will warm a degree or two in the next century. The debate probably will continue for many years as we hear from well-meaning scientists with conflicting data. Frankly, none of us knows for sure whether global warming will occur—I certainly do not. But for the sake of discussion, let us assume that global warming is a significant possibility.

Unlike my colleagues in the environmental movement, I look at the so-called global warming problem from the viewpoint of an electric-utility executive. In particular, my perspective brings with it a very hard sense of economic reality.

Every day we in the utility business must deal with tradeoffs between benefits and costs—not just what we judge to be economically sound but what our customers and regulators will accept as prudent. If the public does not perceive something as beneficial, it will not accept it, no matter how valid the arguments. Thus, unless people believe global warming is actually taking place and poses a personal threat, they will not support drastic or costly measures. However, there are signs that public awareness is increasing. A number of sea-level communities are even beginning to consider the possibility of rising ocean levels in their long-range planning.

We need to examine potential global warming in terms of steps that could be taken quickly and steps that could be implemented only over a long period.

CONSERVATION

In the short run, more efficient use of energy can provide real economic benefits to consumers while helping reduce so-called greenhouse gases. For example, improved insulation of existing buildings is cheap, effective, and fuel-efficient. Better design of new structures also can help. These *reasonable* measures for conserving energy are acceptable to the public, although some of the more extreme measures are not.

The problem with many energy-efficient applications, from the consumer viewpoint, is that they cost money *now* and promise to save money *later.* Relatively affluent people are willing to spend $20 for a light bulb that will save $4 to $5 in electric bills each year. But for most people, a 49-cent conventional light bulb is much more attractive than saving 30 cents a month.

Electric utilities can be a real help to industrial and commercial customers by showing them how they can significantly cut costs by investing in energy-efficient technologies. I am happy to say that in the United States we are making significant progress in this area. But energy conservation measures must be continually reevaluated as circumstances change. Cost-effectiveness of any given measure depends on two externally controlled factors: the cost of energy and the cost of money. If either of those factors changes, all options must be reconsidered.

In the long run, as new technologies become available, they will increase the efficiency of existing conservation applications. Already, prudent use of electricity is having a major effect on electric demand; California's predicted needs for this decade, for instance, have been cut by 12,000 megawatts below 1977 projections.

Conservation can depress the consumption curve, or it can flatten the growth rate. But it is unrealistic to expect that over the long term conservation alone can solve alleged global warming problems.

ELECTROTECHNOLOGIES

On the demand side, a variety of electrotechnologies can improve the total energy efficiency of the economy, provide economic benefits to users, improve local air quality, and reduce production of carbon dioxide. For all those reasons, Southern California Edison is expanding its efforts to inform customers about new electrotechnologies. We recently dedicated a new Customer Technology Applications Center that serves as a focal point for education, communication, and demonstration of clean, energy-efficient electrotechnologies.

In most cases, these technologies give our customers better, cheaper, more efficient ways to run their operations, while meeting the increasingly stringent air-quality regulations of the Los Angeles Basin. For example, there are thousands of furniture makers, metal finishers, and printers in the Los Angeles area who use highly polluting solvent-based coatings. Consequently, air-quality regulators have targeted them for large emissions reductions. Often these manufacturers think they have no alternative but to move their plants elsewhere.

We contact them and invite them to our customer center. When we set up an appointment for a furniture manufacturer, for example, we ask for some samples of uncoated furniture. Our experts—retired furniture manufacturers themselves—take the samples and practice coating them with ultraviolet curing. This method not only saves energy and reduces air pollution but also is much faster—sometimes taking a matter of seconds, rather than hours. Our experts demonstrate the ultraviolet-curing technology to the manufacturer and answer his questions intelligently. As a result, several furniture manufacturers are now considering the use of ultraviolet-cured coatings to meet air-quality regulations.

Another exhibit highlights the benefits of replacing large internal combustion engines with electric motors. Because of a test program just completed, we now know that replacing more than 550 engines would reduce emissions of the oxides of nitrogen in the Los Angeles Basin by more than 30 tons per day.

The dielectric-heating exhibit at our center features radio-frequency dryers and microwave ovens, which can be used in a variety of industrial applications involving heating and drying. These devices are highly efficient and nonpolluting. For example, insulating fiberboards are dried in fifteen minutes with a radio-frequency dryer, as compared with the twenty-four hours needed when using a gas furnace.

The center also has a full-scale model of the "House of the Future," equipped with energy-efficient appliances and computer-controlled cooling, heating, and lighting systems. Here customers can learn how to reduce energy costs without sacrificing comfort or convenience. Other exhibits demonstrate outdoor lighting, home automation systems, and electronic metering.

ELECTRIC TRANSPORTATION

In the world of transportation, people who fear that our society will be forever dependent on petroleum are in for a pleasant surprise. *The* transportation fuel of the future is electricity. There is no cleaner fuel than this. But achieving that future will take planning, commitment of resources, and leadership.

Historically, Southern California Edison played an important role in our region's development of electric rail transport, by supplying inexpensive hydro power to the famous Red Car and Yellow Car systems. Before these systems disappeared from the Los Angeles area in the 1960s, they represented the largest interurban rail network in the United States, operating more than 6,000 daily passenger trains. The trolley systems were Edison's largest single customer.

Today, electric transportation offers southern California a way to clean up its notoriously dirty air *without* surrendering the freedom of movement its residents treasure so much. Gasoline and diesel engines emit a lot of reactive organic gases, carbon monoxide, and nitrogen oxides. But electric vehicles (EVs) emit virtually none.

Even when you consider the emissions from power plants that produce the electricity to charge EV batteries, electric vehicles still produce 97 percent less emissions in southern California than do vehicles with gasoline engines and at least 88 percent less than vehicles with engines using *any* other alternative fuel. In addition, recharging EV batteries would produce 45 percent less carbon dioxide in southern California than gasoline engines currently produce.

Electric transportation covers a wide spectrum of new, evolving, and existing technologies. It means electric trolleys and buses, electric light- and heavy-rail lines, electromagnetic levitation trains, and electric cars, vans, and roadways. I will describe briefly some of these innovations.

In California, light-rail lines operate in San Francisco, Sacramento, San Diego, and San Jose. In our service territory, a line from Long Beach to downtown Los Angeles will begin operation this summer, while a second line to Los Angeles International Airport should begin service two years later. Many more lines are in the planning stages.

Electric heavy-rail transportation is less common but perhaps more important. Bay Area Rapid Transit certainly demonstrated its worth during and after the major 1989 earthquake. The system proved seismically sound and provided a vital service to the Bay Area when the Bay Bridge and the Nimitz Freeway were damaged.

The disaster caused the director of Caltrans, California's department of transportation, to say, "Roads are what we know, but we really have to start emphasizing things we don't know so well." Such as electric mass transit. California needs to develop a balanced transportation system to offset its citizens' tremendous dependence on the automobile.

Engineering problems do not prevent the construction of underground rapid transit in California—but political problems do. Although it is expensive to tunnel each new mile of LA's subway, the cost is only about one-fourth that of each new mile of the city's Century Freeway. And as with other forms of electric transportation, there is almost no air, auditory, or visual pollution associated with subways.

Development of electromagnetic levitation trains, the most rapid form of electric transportation, looks promising. California and Nevada have been studying a private proposal to put a "maglev" train between Las Vegas and Orange County; it is projected to carry about 6.5 million passengers a year by the end of the 1990s, at speeds over 300 miles per hour. The long-range plan is to build a network of maglevs connecting southern California, the Bay Area, Sacramento, Reno, and Las Vegas. Prototype maglevs are operating now in West Germany and Japan.

Electric buses are already here. In San Francisco, buses have operated from overhead trolley wires for years. In 1991 Santa Barbara will have battery-operated electric buses that Southern California Edison will help finance.

Near the University of California's Berkeley campus, a bus is being tested that has its own energy-storage system and an inductive-coupling device to recharge batteries from an electrified track. Berkeley and Caltrans are experimenting with the bus on a 400-foot section of electrified track as part of the Program on Advanced Technology for the Highway. (The same program is investigating guidance systems for vehicles, which may be used in fully automated vehicle highways in the future.) In a few months, a quarter-mile-long electric roadway will be built in Los Angeles at the site of a new real estate development; this will be the "test track" for the same type of electric bus being tried out at Berkeley, as well as for converted electric vans.

Roadway electrification means that cables buried beneath the road surface supply the power to move vehicles. The power is safely transmitted by induction with no physical link between the vehicle and the cables. Electric energy is used both to propel vehicles and to recharge their batteries. Once vehicles leave the electrified roadway, they may move about on battery-stored energy.

If the Los Angeles electric-roadway tests go well, the developer will build electric cables into much of the new community's streets when construction begins in 1992. The vision is gradually to electrify the streets of surrounding areas in West Los Angeles—and eventually to do so in much of the 46,000 miles of roadway in the metropolitan area.

Electrified roadways will eventually make electric vehicles more popular. But already we're working on other ways to popularize EVs. Southern California Edison has purchased fifteen pre-production G-Vans to loan for one-month periods to public and private fleet operators in southern California. We're also lending vehicles to legislators and government agencies in Sacramento and Washington, D.C.

The local South Coast Air Quality Management District (AQMD) is encouraging the use of electric vehicles as well. The district's goal is for 20 percent of the vehicles in our area, or approximately 1.7 million, to be EVs by the year 2000. That figure will rise to 70 percent of the vehicles, or 6.7 million, by 2010. These goals would require sales of 350,000 to 500,000 EVs a year, starting in 1996. Southern California Edison is notifying vehicle manufacturers of this sales requirement in order to bring them into the mainstream of EV development and commercialization.

ELECTRICITY GENERATION

As far as the generation of electricity is concerned, several opportunities exist to reduce the alleged greenhouse effect. First, we can reduce carbon dioxide (CO_2) nationally by using more natural gas and less oil and coal. Electricity generation accounts for about 30 percent of the CO_2 production in the United States, although only about 10 percent in California because of the state's high reliance on natural gas as a primary boiler fuel. For a given electrical output, natural gas creates 30 percent less CO_2 than oil and 50 percent less than coal.

Even though Southern California Edison would like to burn natural gas exclusively, we cannot. In California, there is not enough available natural gas. As a result, we have had to use oil, particularly during winter months. In Nevada and New Mexico, we could not easily switch our power plants from coal, the nation's most abundant and economical fuel, to natural gas. Total reliance on natural gas also would increase the cost of energy. That's because once coal and oil were no longer used for electricity production, natural gas prices would skyrocket.

A second way to counteract a greenhouse effect would be to increase the thermal efficiency of existing fossil-fuel plants. A 1-percent improvement in the average efficiency of Edison's plants would reduce CO_2 production by about 275,000 tons a year. We are testing several techniques to increase efficiency in our steam plants, including steam turbine modifications, use of improved air preheaters, and new burner designs. However, it is not yet clear whether these changes will be cost-effective. And some of the required changes to reduce local emissions of nitrogen oxides in our plants will actually *decrease* the efficiency of these plants, thus *increasing* CO_2 output.

This demonstrates the importance of tradeoffs we often face in making decisions on air quality. In the Los Angeles area, we are fortunate to have the forceful leadership of Jim Lents, executive director of the South Coast Air Quality Management District. Jim understands the economic impacts and tradeoff issues very well. We are working with him and the AQMD to balance air-quality decisions for the greatest public good.

A third approach would be to repower existing boiler-type plants that operate at about 33 percent overall efficiency, which is standard for power plants, with new high-efficiency combined-cycle plants that have an overall 50-percent efficiency level. This gain would reduce CO_2 production from Edison plants by about 4 million tons a year. It is economically feasible if implemented on the basis of replacing existing plants as they wear out.

The fourth and potentially the most significant action we can take to lessen possible global warming is to increase the use of nonfossil energy resources. Southern California Edison has helped develop a wide array of technologically feasible nonfossil means of generating electricity. We have five different solar-energy technologies, geothermal from both "clean" and highly corrosive steam, wind, fuel cells, and hydro. Overall, we have enough experience with these technologies to know they can produce electricity reliably, although not as cost-effectively as more conventional power plants.

Geothermal energy comes the closest to being cost-competitive, but usable geothermal steam supplies exist in very few locations and mostly in the West. The United States Geological Survey estimates as much as 300 trillion megawatt-hours of accessible geothermal resources lie beneath 3.4 million acres of U.S. land, but only about 4 percent occurs as hydrothermal resources—superheated water within permeable rock formations, and water trapped below layers of impermeable rock. Hydrothermal resources continue to provide most of the geothermal electric development.

In southern California, the best source of geothermal energy is in the Coachella Valley, south of the Salton Sea, but we buy geothermal power also from the Mammoth Lakes area in central California, as well as from Nevada and Mexico.

Since the early 1970s, the development of wind technologies for electric power production—aided by state and federal tax incentives—has progressed very rapidly. However, wind energy is not without environmental impacts. Many people consider wind farms unsightly and noisy. And they require extensive road networks that further destroy their natural settings.

Solar power plants that generate steam have provided a successful alternative energy source in the past, but the best hope for real price-performance gains lies with solar cells that convert sunlight directly into electricity. However, the best commercially available solar cells today are only 12-percent efficient and are many times more costly than fossil-fuel plants. In time, solar energy may become a practical solution for those parts of this country and the world where the sun shines intensely and regularly, such as the desert Southwest.

Hydro is a wonderful resource, but a limited one. Most of the prime hydroelectric sites have already been developed, and in North America, utilities and environmentalists disagree about developing the remaining sites. Are we willing to sacrifice our wild rivers in order to avoid burning fossil fuels?

NUCLEAR

Perhaps the most reasonable nonfossil energy resource for generating electricity is nuclear energy. Fusion power will likely not be available until some time in the next century. So let's look at the current technology of fission reactors.

Fission reactors must be considered an important part of our overall global warming strategy. Nuclear development, however, has stagnated in the United States because of public concerns over reactor safety, waste disposal, and costs—even though fission has proved itself a safe, reliable, and nonpolluting source of energy. This statement may seem suspect after Three Mile Island and Chernobyl. But in fact, in most of the world, including the United States, reactors are very safe. Since Three Mile Island, we have retrofitted our nuclear plants with redundant safety systems. If one safety system were to fail, another would automatically take its place. An accident such as that at Chernobyl would be extremely unlikely outside the Soviet Union because of much stricter safety standards and procedures elsewhere.

There are some who may worry about nuclear power in the hands of terrorists or radical Third World countries. In fact, commercial nuclear energy poses no threat: it is more complicated to use the by-products of a nuclear power plant for manufacturing weapons than it is to produce them by other means.

The attitude toward nuclear energy is much more positive in the rest of the world than in the United States. France, for example, derives 70 percent of its electricity from nuclear power plants; Belgium, 66 percent. Bong Sum Lee, South Korea's minister of energy and resources, says nuclear power is the solution to environmental and supply problems—especially in resource-poor developing countries.

But the United States derives only 18 percent of its electricity from nuclear power. When you consider that 8 pounds of uranium can make as much energy as 12 million pounds of coal—without polluting the air—it makes sense to put politics aside and move prudently forward with nuclear power in the United States. However, public acceptance of a new generation of fission reactors in the United States will require major changes:

First, we must develop a *single, standardized design* for each basic reactor technology. Today we know how to design and build what is referred to as a "passively safe" reactor—that is, one that will shut itself down if anything goes wrong, without a need for operator intervention. Paul E. Gray, president of the Massachusetts Institute of Technology, recently described such a plant in *The Wall Street Journal.* He said it would need less complex, less costly safety systems than current reactors and therefore would be much more cost-effective.

Second, all utilities must be required to use the same standardized designs, so consumers can obtain the economic, operational, and safety benefits that result from such standardization.

Third, a simplified licensing and permitting process is needed to certify new reactor sites automatically if they meet established criteria.

Finally, solutions must be found to the technical and political problems surrounding disposal of high-level radioactive waste—spent fuel rods. This is primarily a political issue: nobody wants a disposal site in his backyard. Government and utilities need to get on with the process of educating the public about disposal sites and then selecting them, and the public must be involved in the selection process. This is a national policy issue—one that balances local interests against national interests. I hope that when we finally begin permanent disposal of high-level waste, we do so in a way that will make the waste recoverable by future generations. After all, what we consider an unwanted by-product of electricity production still contains 75 to 90 percent of its original energy.

A strong electric power system is imperative. Today there are about 100 nuclear power plants operating in the United States. More electric capacity will be needed, because our population is growing and improved electrotechnologies will replace less efficient, more polluting processes. The need for kilowatt-hours will grow in the next few years despite the strongest of conservation efforts.

Even without an increase in demand, many existing power plants will need to be replaced within the next several decades. The average age of a power plant in the Southern California Edison system is twenty-five years; that is the national average as well. What kind of power plants will we build to replace them?

SUMMARY

There is no free lunch at the energy table. In dealing with the complex give-and-take that affects the electric utilities, I am reminded of a fable: A dog with a bone in its mouth comes to a pond and sees its reflection in the water. It thinks there is another dog there, with a bone in its mouth, and tries to get that one too. In doing so, it drops its bone in the water and winds up with nothing.

The point is that no one side can have it all. When it comes to an issue as vital as protecting the environment, tradeoffs between benefits and costs must be weighed with the utmost judiciousness.

Likewise, every fuel involves tradeoffs. Alternatives and renewables— such as solar, wind, biomass, and geothermal energy—are growing in use. But because the sun does not shine, the wind does not blow, and the brine does not boil everywhere or often enough, these technologies will remain niche applications.

Fossil fuels and nuclear power will provide most of our base-load capacity. Coal is the most abundant generating resource in the United States. But as presently used, it is highly polluting. Developments in clean-coal technology are helping to resolve this problem.

We are potentially vulnerable as far as supplies of oil and gas are concerned. They are exhaustible sources and as such will not always be available, especially at competitive prices.

Nuclear energy has two strong advantages—it is clean, and it can help us gain energy independence as a nation. Yet I doubt there's anyone willing to build nuclear power plants like the ones already on line. The "first-generation" nuclear plants currently operating in the United States were designed and constructed to a set of criteria different from what would be applied now. We must ask ourselves what must we do today to ensure maximum use of nuclear power as a generating resource tomorrow. What safety standards will it take to make nuclear power acceptable to the public?

Improvement and change are possible, but not without a strong national commitment to nuclear energy. I hope growing interest in the possible problems of global warming will help us develop that national commitment.

We have no easy quick-fixes or single-shot solutions to global warming. But we have at our disposal today a number of strategies to reduce energy consumption—and to reduce the global warming impact of the energy we produce. I think that is basically good news for us and the world.

WILLIAM W. KELLOGG

Theory of Climate:
Transition from
Academic Challenge
to Global Imperative

Dr. William W. Kellogg served for almost ten years as director of the Laboratory of Atmospheric Sciences at the National Center for Atmospheric Research, and as a senior scientist there until his retirement in 1987. He was a member of a number of government advisory committees, including the President's Science Advisory Committee's Panel on the Environment, the U.S. Air Force Scientific Advisory Board, and the NASA Space Program Advisory Council. He is a past president of the American Meteorological Society.

In 1985 Dr. Kellogg received a commemorative medal from the Soviet Geophysical Committee and a certificate of appreciation from the U.S. Department of Commerce. He is coauthor of *The Atmospheres of Mars and Venus* (with Carl Sagan) and *Climate Changes and Society* (with Robert Schware), and has written many reports and articles. Dr. Kellogg's research has focused on the dynamics of the upper atmosphere, the atmospheres of Mars and Venus, the prediction of radioactive fallout, development and use of meteorological satellites, and most recently, the influence of human activities on global and regional climate, and the societal effects of climate change.

The concept of the greenhouse effect goes back over 200 years. The fact that the Earth's surface has always been warm enough to keep the oceans from freezing and therefore has permitted life to form and evolve is due to a few trace gases in the atmosphere that trap some of the outgoing infrared radiation that would otherwise escape to space. Now mankind is adding more of these greenhouse gases to the atmosphere, and that addition is turning up the global thermostat—and gradually changing many other things on the Earth. We are fairly certain this is indeed happening. Policymakers and scientists are trying to figure out what, if anything, to do about it. If we are to consider climate change when we make long-range plans, it is essential that we understand the likely patterns of future change as well as possible. Who will be the winners, who the losers? And then there is a difficult value judgment: To what extent is climate change unacceptable? What are the countries of the world willing to sacrifice to avert it?

A BRIEF EARLY HISTORY

In the past few years, thanks to headlines and media hype, "greenhouse effect" has almost become a household term. And it seems to have taken on a sinister meaning, since it symbolizes another of the ways in which mankind is afflicting planet Earth. Many of us worry about the environmental degradation we see around us and read about in the papers, and we have a justifiable sense of tribal guilt.

Although the environmental ethic so widely embraced lately is a creation of the late twentieth century, its seeds go back hundreds of years. As I will show, the realization that mankind can change the climate of the Earth, which has gained general credence since about 1970, is actually not a new one (Kellogg, 1987; Revelle, 1985).

The first person to propose formally that gases in the atmosphere could absorb some of the "heat radiation" the Earth's surface is constantly emitting was a Frenchman, Jean-Baptiste Joseph Fourier (1768–1830). He suggested that the Earth is kept warm by this process, in the same way the glass of a greenhouse keeps the interior warm on a cold day, and he called it *"l'effet de verre"* (the glass effect). It seems he misunderstood the mode of action of a greenhouse, but he nevertheless demonstrated a remarkable early insight.

In the following decades, other scientists filled in a great many of the pieces in the puzzle of the greenhouse effect. For example, John Tyndall (1820–1893) actually measured the absorption of infrared ("heat") radiation by carbon dioxide and water vapor, and he showed conclusively that the presence of these atmospheric constituents could significantly raise the Earth's surface temperature. He was apparently the first to make an important additional deduction, namely that glacial periods may have been caused by a decrease in atmospheric carbon dioxide (Tyndall, 1863); this hypothesis recently has been verified for the most part.

Just before the turn of the nineteenth century, two remarkable studies were published that advanced greatly our realization of the effects of carbon dioxide on climate. It was already recognized that the concentration of carbon dioxide was probably increasing as mankind took carbon out of the Earth, in the form of coal, petroleum, and natural gas (fossil fuels), and burned it. To what extent could this change our climate?

Svante Arrhenius (1859–1927), a Swedish chemist who received the Nobel Prize for Chemistry in 1903, attempted to calculate the effect on the Earth's average temperature if the concentration of carbon dioxide were to be doubled (or halved). This was a truly brilliant piece of research, and in those days there were no computers (except people), and information about the infrared absorption characteristics of the trace gases in the atmosphere was rather primitive. For his estimates of atmospheric absorption due to carbon dioxide and water vapor, Arrhenius used measurements obtained by American Samuel Langley of infrared radiation from the moon as it passed through the atmosphere at different angles above the horizon and at different humidities. He combined these with independent measures of carbon dioxide to estimate its current optical depth, or ability to absorb radiation, and then calculated that doubling the carbon dioxide would raise the average surface temperature by 5° to 6° Kelvin, the larger warming occurring at higher latitudes. (In a book published twelve years after the research, Arrhenius cited an average temperature rise of 4°K.) A decrease of carbon dioxide to two-thirds of the then present concentration would cause a cooling of 3° to 3.4°K, he added. In these calculations he took into account the likely increase of total water-vapor content with increasing temperature, an important "feedback mechanism" (that term would not be invented until some fifty years after his research). On the basis of his calculations, Arrhenius deduced correctly that "if the quantity of carbonic acid [carbon dioxide] increases in geometric progression, the augmentation of the temperature will increase nearly in arithmetic progression"; if a doubling causes a 4°K increase, for instance, a quadrupling would cause an 8°K increase (Arrhenius, 1896, 1908). In discussing the implications of what is happening now to our climate we will have occasion to quote again from this great scientist.

An American geologist, Thomas C. Chamberlin (who became president of the University of Wisconsin), built on the insights provided by Arrhenius to "frame a working hypothesis of the cause of glacial epochs on an atmospheric basis" (Chamberlin, 1899). Much of his argument deals with the changes in sea level relative to the continents, and the effects these changes would have on the weathering of silicate rocks and on atmospheric carbon dioxide depletion, a subject also treated by Arrhenius. Chamberlin suggested a mechanism involving the effects of sea level on continental ice sheets, and he showed, among other things, that this would result in the cyclical behavior of glaciations and interglacial periods in the last million years or so as the balance between carbon dioxide removal by weathering of rocks and replenishment by volcanic emissions was altered. (An alternative theory based on cyclical perturbations of climate due to changes in the Earth's orbit and axis of rotation was suggested by N. Milankovitch (1930) some thirty years later, but nevertheless Chamberlin's hypothesis regarding carbon dioxide is a reasonable explanation of cyclical climatic variations throughout geologic time, and is still believed to be relevant.)

It is interesting to note that an important aspect of this hypothesis, namely the role the oceans play as a major reservoir of carbon dioxide, was recognized by Chamberlin and studied by one of his students, C. F. Tolman, Jr. (1899). Although relatively little was known then about the exchange rates of water masses within the oceans, and their influence on storage times of carbon dioxide, the early insight provided by Tolman was significant. As we shall see, the oceans are a major player in the drama of climate change.

As early as the beginning of the twentieth century, then, several of the principle building blocks of a theory of climate change were already in place. Obviously there were serious gaps in our knowledge of the planetary system that determines climate—and there still are—but enough was known to suggest strongly that mankind could alter its delicate balance.

What could hardly have been understood at the turn of the century was society's insatiable appetite for energy and the following decades' rate of increase in consumption of fossil fuels. From that time on, until 1973, the consumption of fossil fuels was to increase at a phenomenal 4 percent per year, a rate corresponding to a doubling time of about fifteen years (Rotty and Marland, 1986). Who would have believed that back in 1900?

The writings of Arrhenius were not entirely forgotten, but they seem to have had little impact on our thinking for a long time, until the 1970s. On the whole, except for a few discerning and concerned individuals, the scientific community and the public seemed largely oblivious to the possibility that mankind could influence the greenhouse effect and hence the temperature of the Earth.

Among those who did become concerned was the English atmospheric chemist G. S. Callendar, who made extensive measurements of carbon dioxide concentrations over a period of years and who reevaluated some of the earlier measurements. Although the analytical techniques he used were not nearly as accurate as modern ones, Callendar was able to show there did indeed seem to be a slow rise of the concentration of carbon dioxide, in line with the belief that burning of fossil fuels could be the cause (Callendar, 1958).

With better knowledge of how trace gases in the atmosphere—water vapor, carbon dioxide, methane, nitrogen oxides, chlorofluorocarbons (CFCs), and so forth—absorbed infrared radiation passing through the atmosphere, it became possible to improve on the earlier calculations of Arrhenius (Ramanathan, Singh, Cicerone, and Kiehl, 1985). Unfortunately, knowing the absorption spectrum is not enough to calculate the total greenhouse effect, since there are many other factors related to climate that have to be taken into account. The distinguished German scientist Fritz Möller, for example, published a calculation (1963) that correctly evaluated the purely radiative part of the problem but made a questionable assumption about the way cloudiness would change as the world warmed. The result was far too large a warming compared to either Arrhenius's findings or our modern climate models.

By the 1950s, a larger community had begun to recognize and address the important connection between carbon dioxide and climate. John Von Neumann (1955) wrote about the possibility of "climate control." In 1957, Roger Revelle and Hans Suess, two scientists at the Scripps Institution of Oceanography, made a statement in the journal *Tellus* repeated many times since: "Human beings are now carrying out a large-scale geophysical experiment," namely, testing the greenhouse effect of carbon dioxide by actually changing its concentration. Revelle and Suess also pointed out that the newly added carbon dioxide would probably remain in the atmosphere for many centuries because of the slowness with which the oceans could absorb it.

In the period after the International Geophysical Year (1957–1958) there was a growing general awareness of what might be taking place as a result of our "geophysical experiment." In 1963, the Conservation Foundation sponsored a meeting, and its report stated the situation more succinctly than anyone had before: "It is estimated that a doubling of the carbon dioxide content of the atmosphere would produce a temperature rise of 3.8 degrees [Celsius]"—the time scale involved is left unspecified, however (Conservation Foundation, 1963).

Just two years later, in 1965, the august President's Science Advisory Committee (PSAC) published under the seal of the White House a report of its Environmental Pollution Panel titled "Restoring the Quality of Our Environment" (PSAC, 1965). (The panel's chairman was John Tukey, and Roger Revelle was chairman of the Subpanel on Atmospheric Carbon Dioxide.) The PSAC report deals with many aspects of air and water pollution, but it will also be remembered as a first public recognition by the U.S. government that climate change could be caused by human activities and that this would have important consequences for the world.

EMERGENCE OF A QUANTITATIVE THEORY OF CLIMATE

This obvious fact has been mentioned already: The planetary system that determines the climate of each part of the Earth is exceedingly complex. There are so many interconnected components, as suggested in Figure 1, and it is an insuperable task to take them all into account in a theoretical "model." The best we can hope to do is to deal with the more important factors (Ramanathan, 1988; Schneider, 1989; SMIC, 1971).

The dilemma of Earth's complexity is still a challenge, but we have made progress since Arrhenius. He had a theoretical model in mind when he made his historical calculations almost a century ago, but it seemed that little further progress had been made to improve on it until S. Manabe and R. T. Wetherald published another calculation (1967) of the effect doubling carbon dioxide would have on temperature, this time with more or less modern data on the geophysical interactions—and the assistance of computers. The model was quite simple, the calculation done along one vertical line, as if the atmosphere could be considered the same everywhere on the globe. Manabe and Wetherwald took the change of water vapor into account in a way similar to the approach of Arrhenius, but with refinements; essentially, they kept the relative humidity constant as the atmosphere warmed. They also showed how assumptions about changes in cloud distribution and

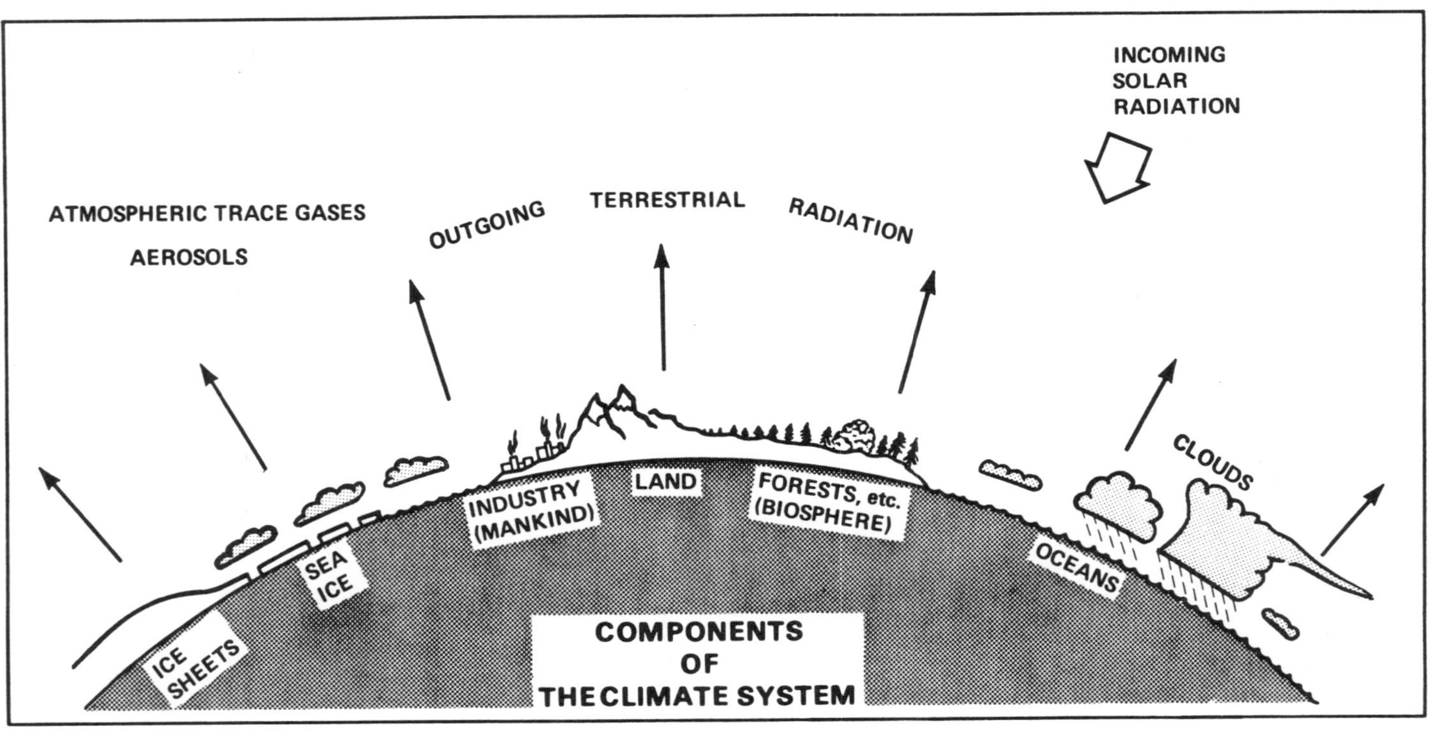

FIGURE 1. These components of the Earth's climate system interact with each other and together determine our climate.

temperature would affect the results. Their conclusion: A doubling of carbon dioxide concentration would probably result in approximately a 3°K average warming—not far from the result found seventy years earlier.

Progress in science is a slow and tortuous process, and in the gradual development of a rational and quantitative theory governing our climate system it would be hard to identify when the "breakthroughs" happened. However, I will venture the opinion that a most important new thrust occurred in 1969, with publications by two scientists on opposite sides of the Atlantic. They were Mikhail Budyko at the Main Geophysical Observatory in Leningrad (Budyko, 1969) and William Sellers at the University of Arizona in Tucson (Sellers, 1969).

The two models proposed were quite different in detail, but both took a leap forward by showing that one could take into account the basic fact that heat is added to the atmosphere at the equator by an excess of absorption of sunlight, and removed in the polar regions by an excess of infrared radiation back to space. The atmosphere acts as a great heat engine, and it is just that difference between the equatorial heating and the polar cooling that drives the atmospheric circulation—the analogy has often been made to an engine's boiler and condenser. These two new models took large-scale atmospheric circulations into account, albeit quite crudely, and showed how they transported thermal energy from equator to pole. They also showed how the snow line would change in latitude as temperature changed. This kind of model is called an "energy balance model" to distinguish it from Manabe and Wetherald's "convective equilibrium model."

Now new kinds of experiments could be performed. Sellers, for instance, showed how the model temperature would alter with changing solar radiation, and Budyko showed that even a slight decrease in heating from the sun would cause the snow line to move to the equator, and thus the Earth would become permanently frozen over. These results could be interpreted also in terms of changes in carbon dioxide, and the new results turned out to be similar to the earlier ones.

Obviously, a great deal of progress has been made in climate models since 1969. Supercomputers allow "number-crunching" programs to run in a shorter time than was conceivable a couple of decades ago. As a result (as will be explained shortly), the world can be treated as a three-dimensional whole. Satellites have given meteorologists and oceanographers a view of the Earth never possible before, and we can now compare weather maps drawn by our computer models with maps of "the real thing." Last but not least, governmental support has been given to the teams of scientists needed to develop these incredibly sophisticated computer programs and provide the physical insights needed to make them as realistic as computer speed and human ingenuity allow.

The most advanced climate models consider not only the atmosphere but also the dynamic oceans, and they calculate both the effects of the land with its mountain ranges and shifting vegetation patterns, and the distributions of clouds and precipitation (still a most difficult factor to deal with). Furthermore, the models track snow on land and ice on the ocean, and even calculate soil moisture, which is the net result of precipitation, evaporation, and runoff to rivers. All of that complexity (actually much more than I have described) seems almost incomprehensible, and some of the modelers have been known to remark ruefully that their models are becoming as complicated and hard to understand as the climate system itself. And still there are many details to be added before these marvels of computer science can reliably simulate planet Earth (e.g., Ramanathan, 1988; Schneider, 1988, 1989).

A number of research organizations are developing climate models, but the most active are the National Center for Atmospheric Research (NCAR), the National Oceanographic and Atmospheric Administration's Geophysical Fluid Dynamics Laboratory (GFDL), the National Aeronautics and Space Administration's Goddard Institute for Space Science (GISS), Oregon State University (OSU), and the United Kingdom Meteorological Office (UKMO). While the various models differ from each other, all of the above organizations have successfully linked general circulation models of the atmosphere (GCMs) to dynamic oceans that respond to wind-stress, temperature, and salinity gradients.

These climate models divide the globe into a grid, each element of the grid representing about 400 to 1,000 kilometers on a side (this varies between models); the grid is then duplicated at a number of levels in the vertical (typically around ten; the OSU model uses only two). This creates a three-dimensional network of grid boxes—10,000 to 20,000 of them, depending on the model. To bring the system alive, as it were, the computer goes from grid box to grid box and integrates the seven time-dependent equations governing momentum, heat, and conservation of mass, starting with the condition arrived at in the previous integration. These time steps are taken at fifteen- to twenty-minute intervals in model time for the atmosphere, but much less often for the more slowly responding ocean. Each iteration moves the system along in a physically coherent and consistent way, like the frames of a movie film. Indeed, the output of such a model run on a computer is a progressive sequence of surface and upper-air maps much like our weather maps.

Modern weather forecasters use a version of the GCM to make their daily forecasts. However, instead of integrating the atmosphere (together with the oceans) for a period of a few days, as is done for a weather forecast, the climate models are typically run for twenty to forty years of model time to obtain a representative set of climate statistics. To get an idea of what such a massive computation costs, it takes about ten hours of time on the Cray supercomputer to integrate the NCAR model for just one year (Schneider, 1988). That may be one explanation why there are so few organizations active in the climate-modeling business.

REGIONAL CHANGES DUE TO A GLOBAL WARMING

Until the three-dimensional atmosphere-and-ocean models were developed, there was relatively little that could be said about regional changes that would accompany a global warming. Climatologists expect the changes will vary enormously from place to place, and the shifts of precipitation could be very significant.

However, as has been pointed out, climate models must divide the globe up into a grid whose horizontal resolution is 400 to 1,000 kilometers on a side, and little can be said about finer details. The smaller-scale processes inside a grid box are treated as aggregates, or averaged, in the model calculation. There will be no alternative until computers are powerful enough to deal with higher-resolution grids—and there will, of course, always be a limit to how far we can shrink the grid spacing.

Since these marvelous models are the best analytical tools we have for looking at the warmer climate of the future, we must learn what we can from them, imperfect as they may be. As will be shown, we can supplement the model results with lessons from past warm periods. This can serve as both a check on the models and a measure of our success.

A visitor to NCAR from Peking University, Zong-ci Zhao, and I were curious to see just how well the five state-of-the-art climate models mentioned above agreed with each other in their predictions of changes in soil moisture when carbon dioxide was doubled. We chose soil moisture rather than temperature or another aspect of weather and climate because it is crucial in determining where things can grow, and because it is a difficult variable to calculate and therefore presents a good test. Recall that soil moisture is the result of at least three factors: precipitation, evaporation, and water runoff when the ground is saturated. To complicate matters further, snow on land does not contribute to soil moisture until it melts; the way meltwater in the spring is handled turns out to be a delicate matter and creates differences between model results (Meehl and Washington, 1988).

When Zhao and I first looked at the maps of soil moisture change they appeared to be rather different, but when we laid them on top of each other we found there were places of agreement. It would have been surprising if there had been perfect agreement, but we thought it significant that they gave the same picture over large areas.

The results of our analysis are shown in Figure 2 for North America and Figure 3 for the region dominated by the Asian monsoon, including Zong-ci Zhao's home (Kellogg and Zhao, 1988; Zhao and Kellogg, 1988). For North America the picture seems to be a general increase in soil moisture in winter in the north and a drying-out in Mexico. In summer the models call for a drying-out over most of the continent, with the exception of areas along the Gulf and Pacific coasts. In Asia the pattern of change seems merely to augment the dominance of the monsoon, with summertime rains in the south and dry conditions in the north, and in winter a kind of mirror image of the same changes.

Is there a way to check whether these model results are reasonable? Indeed there is, since we can ask whether the model changes are consistent with changes that actually took place during anomalously warm periods in this century or the more distant past. For North America the changes are entirely consistent with rainfall deviations during the warmest summers in this century, that is, a drying-out in the Midwest but wetter conditions along the Gulf and Pacific coasts (Jäger and Kellogg, 1983); there is less agreement in winter. Good agreement results when we compare the model predictions with conditions during the warm Altithermal period of 4,000 to 8,000 years ago, that is, dryness in the Midwest but wetter along the Gulf and Pacific coasts. As for the Asian monsoon region, it is well known that the Rajasthan desert of northwest India was a prosperous agricultural community during the Altithermal period, and there are ruins of large cities there, now covered by sand (Bryson and Baerreis, 1967).

In Figure 4 are sketched the regions of the world that may be destined to become drier or wetter. We do not have a great deal of confidence in this picture, except for the areas indicated with dashed lines, but it can be considered a "best guess" based both on climate modeling and on the history of past warm periods (Kellogg, 1982; Kellogg and Schware, 1981). Note, for example, that North Africa should become wetter. During the warm Altithermal period, the Sahara was not a desert but a savannah supporting nomads and their cattle, and at that time the valley of the Tigris

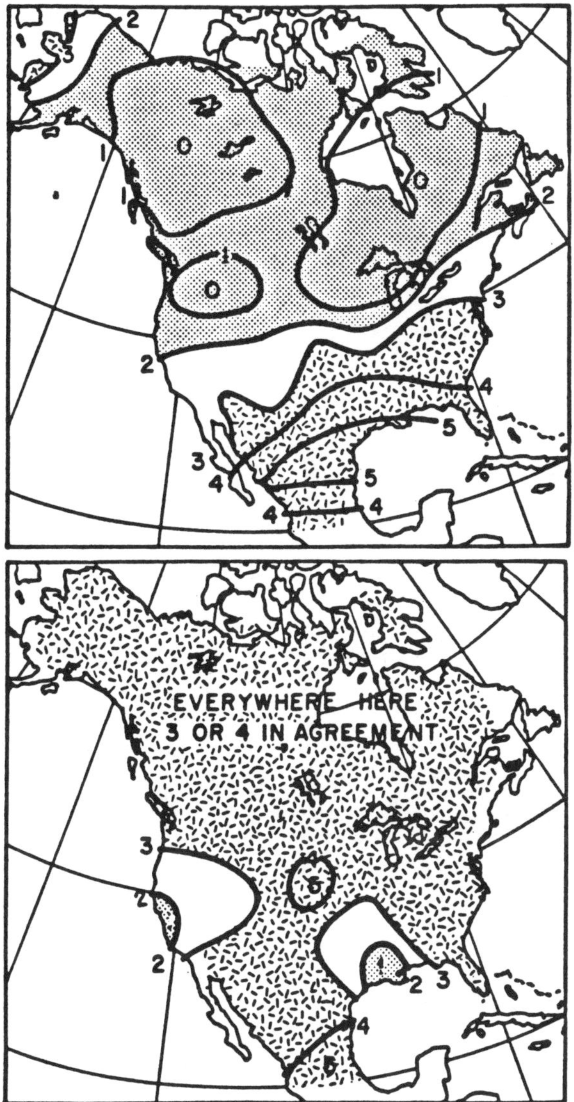

FIGURE 2. Maps of North America showing agreement (or disagreement) among the five climate models named in the text on the direction of soil moisture change when the amount of carbon dioxide in the atmosphere is doubled. Map (a) is for winter, (b) is for summer. Areas shaded with broken lines represent agreement among three or more of the models that there would be a decrease of soil moisture on a warmer Earth; areas with small dots represent agreement among three or more that moisture would show an increase. (*Source:* Kellogg and Zhao, 1988.)

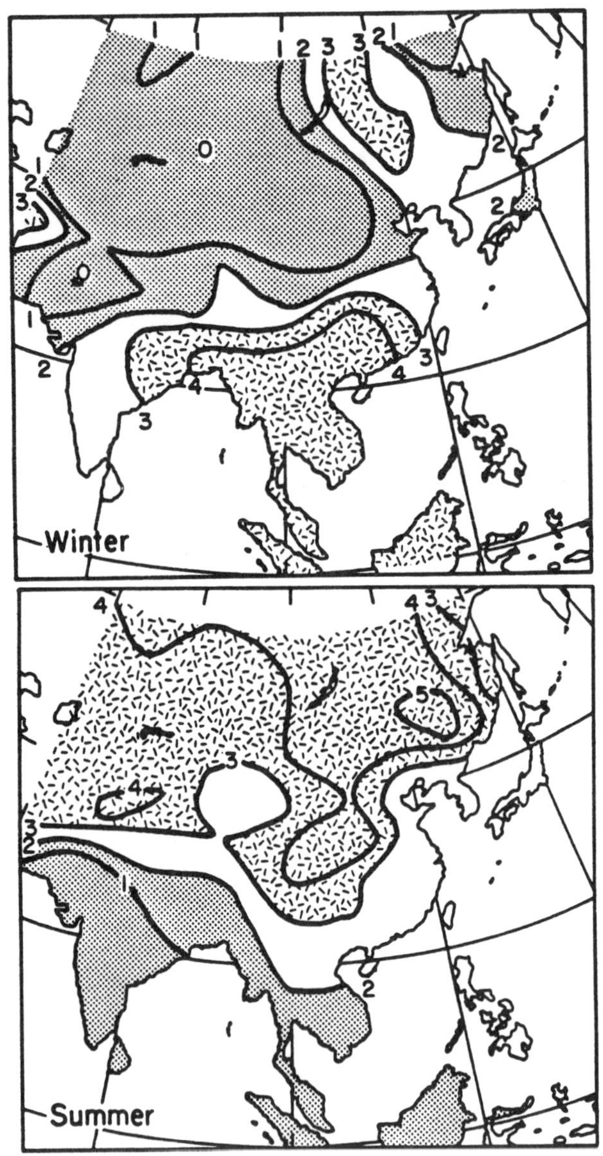

FIGURE 3. Maps of the region dominated by Asian monsoon circulation, indicating information corresponding to that in Figure 2. (*Source:* Zhao and Kellogg, 1988.)

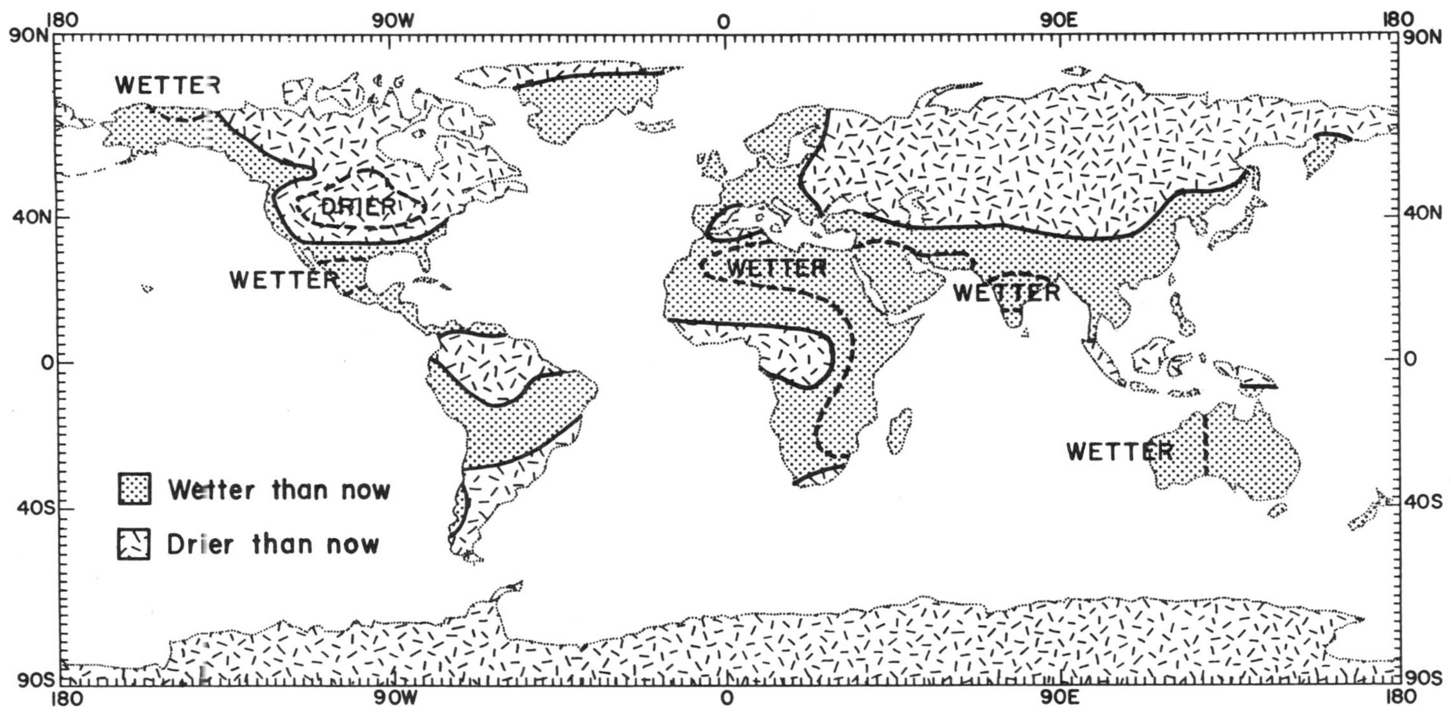

FIGURE 4. This map represents a "best guess" or scenario of possible soil moisture patterns on a warmer Earth in terms of regions where it may be wetter or drier than now in summer; in the tropics, however, patterns probably apply to the rainy season. The map is based on climate model experiments and reconstructions of past conditions, as explained in the text. Broken lines and labels indicate agreement among the various approaches of study.

and the Euphrates could indeed be called "the Fertile Crescent."

Such findings by climate modelers and paleoclimatologists have attracted considerable attention lately, since they seem to suggest where the winners and the losers will be found as the world grows warmer. However, it is recognized that these findings are still tentative, and few politicians are willing to take them into account seriously when making policy decisions. Everyone is waiting for climatologists to do more homework and draw a more credible and definitive picture of the future. We are working on it!

WHAT IS TO BE DONE?

Other essays in this book deal with our responses to the question of global change, and they are written by astute observers of the political and governmental scene. As a physical scientist I cannot claim to be an expert on the ways of humanity, but there may be a place for a point of view that concentrates on a few inescapable verities.

The most obvious is that the global climate change we anticipate, which apparently is already starting to take place, is caused by humanity and its insatiable appetite for energy. The cheapest and most convenient form of energy generation has been the burning of fossil fuels, and there is every indication now that our use of these fuels will continue to increase in the future, quite possibly at an ongoing escalation rate of 2 percent per year. Even if industrialized countries are able to constrain their reliance on fossil fuels (and thus limit the consequent release of carbon dioxide), as seems within their power now (Reinhold, 1989), developing countries will want to continue to exploit this most economic energy source. Who are we to deny them that opportunity? We in the industrialized world are the 25 percent of the human population currently generating more than 75 percent of the energy.

The time has arrived for us to face agonizing policy decisions. These decisions are displayed in Figure 5 in the form of a "decision tree," with the take-charge activists who want to control climate change on the leftmost branch, and those satisfied to muddle through on the rightmost branch. As I see it, where one stands—on the right or the left or somewhere in between—depends on a value judgment about how serious the global warming will be. As we know, we cannot be sure about what will happen, and our crystal ball's predictions of the future climate are still cloudy. And so there are advocates for action and advocates for nonaction.

Let us sample a couple of opinions expressing the happy thought that a

Perception of future environmental change—choice of:

Preventing (averting) or delaying the change		Adapting to the change, assumed to be inevitable	
Purposeful intervention (Take charge)	Counting on providence (Wait and see)	Purposeful mitigation (Genesis)	Ad hoc adjustment (Muddle through)

FIGURE 5. A decision tree showing the range of policy choices that could in principle be made in order to cope with climate change. The actions on the left would presumably have to be taken internationally and on a worldwide basis (see text), but those on the right can be taken at any level of society.

global warming will be good for humanity and planet Earth. In his famous book *Worlds in the Making* (1908), Svante Arrhenius speaks of a future global warming:

> We may find a kind of consolation in the consideration that here, as in every other case, there is good mixed with the evil. By the influence of the increasing percentage of carbonic acid [carbon dioxide] in the atmosphere, we may hope to enjoy ages with more equable and better climates, especially as regards the colder regions of the earth, ages when the earth will bring forth much more abundant crops than at present, for the benefit of rapidly propagating mankind.

And here is what the Soviet climatologist Mikhail Budyko (whose name has been mentioned already as one of the pioneers of climate theory) told an international meeting in Hamburg in November 1988 (Budyko and Sedunev, 1988):

> Although the available information shows that moisture conditions in a number of regions have already deteriorated [because of] global warming, it is probable that beginning with the first quarter of the twenty-first century the moisture conditions will improve everywhere. This casts doubt on the expediency of carrying out very expensive actions aimed at retarding or terminating global warming during the nearest decades.

Now that we've heard from the optimists, here is an excerpt from the report of a distinguished group of scientists and policymakers who met in Villach, Austria, and then in Bellagio, Italy, in fall 1987 (Villach/Bellagio,

1988). The meeting was organized by the World Meteorological Organization and the United Nations Environment Programme. The report summarizes in some detail the best scientific opinion about the greenhouse effect and its impacts, and then states the opinion of the conference, as follows:

> The rate of global temperature change that would occur if current trends of GHG [greenhouse gas] emissions were to continue [is] large compared with observed historic changes and would have major effects on ecosystems and society. For this reason a coordinated international response will become inevitable.
>
> Adaptation strategies for responding to a changing climate adjust the environment or our ways of using it to reduce the consequences of a changing climate. Limitation strategies control or stop the growth of GHG concentrations and limit the climatic change. A prudent response to climatic change would consider limitation *and* adaptation strategies.

After the Greenhouse/Glasnost meeting held in Sundance, Utah, in August 1989, the two organizers, Robert Redford and Roald Sagdeev, wrote a joint letter to the Soviet Communist Party Chairman Mikhail Gorbachev and President George Bush conveying the message of the meeting (Greenhouse/Glasnost, 1989). Their letter, which reflected the sentiments of a sophisticated group of scientists, statesmen, journalists, and prominent media figures that became the basis for this book, stated in part:

> The USA and the USSR are the two largest producers of greenhouse gases. The USSR and the USA are major sources of the world's scientific knowledge which can be deployed to restrain global greenhouse emissions. Therefore, we have a responsibility to provide leadership in the search for common solutions to the global warming problem, and for the environmental security of the world. In this search, we fully recognize the necessity of integrating sustainable development of the Third World. . . . We believe the struggle for prevention of greenhouse gas emissions should be raised to the level of international security.

Among the industrial countries there are still fervent differences of opinion about what, if anything, is to be sacrificed to preserve our climate. High-level government meetings have been held to discuss such policy matters, most recently in Noordwijk, the Netherlands, in early November 1989. This Ministerial Conference on Atmospheric Pollution and Climate Change was attended by seventy governmental delegations. The U.S. delegation was led by William K. Reilly, head of the Environmental Protection

Agency, and several heads of state represented other countries. There seemed to be little disagreement over the seriousness of the global environmental crisis. The proposal that gained the most general favor in early discussions was an initiative calling for industrialized nations to freeze carbon emissions at 1988 levels by the year 2005. While there appeared to be general support for this idea, it did not become a part of the conference's final declaration. Four economic superpowers, the United States, the USSR, the United Kingdom, and Japan, refused to agree to such a restriction (Ralof, 1989).

It is not hard to fathom the reasons for a government's reluctance to promise it will limit the use of fossil fuels. In both the United States and USSR enormous reserves of coal and other fossil fuels exist, ready for exploitation. There, as in most industrialized nations, the corporations and ministries that control production, distribution, and use of fossil fuels are among the most powerful in the world, and are unlikely to welcome imposed constraints, either from home governments or the global community. Furthermore, the less developed countries, some of which—like China—also have large deposits of fossil fuels, are struggling to become more industrialized, and are forced to resort to the cheapest and most convenient available energy source, fossil fuels (Kellogg and Schware, 1981, 1982; Schneider, 1989).

On the other hand, the Noordwijk proposal almost passed did not seem to constitute a very drastic cutback on carbon dioxide production, at least for the United States and other wealthy nations (Reinhold, 1989). The United States is presently one of the most wasteful users of fossil fuel energy in the world, vying with East Germany for that dubious distinction, with a conspicuous consumption of about 5 tons per capita per year (1986 figures). West Germany and Japan, among others, use less than half that much per capita, yet have equally high living standards (Marland, 1989).

The most obvious path to follow in an effort to cut back on carbon dioxide emissions is to conserve energy. Industrialized countries are indeed making some progress in this respect, and most have leveled off or even decreased their rate of increase in the use of fossil fuels (except the United States's automotive and electrical-generation sectors) (MacKenzie, 1988). There are a variety of sensible suggestions for encouraging reduction, such as higher gasoline taxes and penalties for driving large gas-guzzling cars; charges to factories or utilities for carbon emissions; programs to introduce more efficient heating, lighting, and refrigeration equipment; better insulation of buildings; and so forth.

The biggest step forward will be taken when alternative, nonfossil energy sources become economically competitive. Solar, biomass (a form of solar, in a sense), and wind energy are all gaining acceptance in many countries, including the United States. It is undoubtedly time to have another hard look at nuclear energy too, since there are some new developments in inherently safe reactors that may make them more acceptable to the public as well as more cost-effective (Weinberg, 1989).

The final and most basic verity in this mix of worldwide societal and environmental problems is the seemingly inexorable growth of the human population. It has been pointed out repeatedly that the biggest population explosions are in the poorer countries of the world, the very countries that even now can least afford to clothe and feed their people. In one sense this is a separate issue from that of fossil fuels and climate change, but in a larger sense it is very much related. Our demand for energy is driven by the demands of the people for it; and the climate change that may be in store for the world will often strike hardest at marginal areas, places already struggling to feed their populations. So the control of the continued growth of the human population should be a number-one priority—even though we see serious difficulties in overcoming the political and religiously inspired opposition to measures for achieving it. In the long run, success here will do the most to reduce human suffering in the poorer parts of the world.

Will we, all mankind, find a way to deal with these problems, never faced on a global scale before? I refer you back to Figure 5: it seems to me that the "genesis strategy" (a term coined by Stephen Schneider) is the course to follow, at least until the community of nations decides to "take charge"— and that may not be for many years. To the extent we accept the proposition that a global warming is inevitable, we can seize now on opportunities to deal with it, to mitigate its more serious adverse consequences, and to take advantage of the favorable aspects of the change where it is possible to do so. Arrhenius was probably right when he said, "As in every other case, there is good mixed with the evil."

REFERENCES

Arrhenius, Svante, "On the Influence of Carbonic Acid in the Air upon the Temperature of the Ground," *Philisophical Magazine* 41 (1896), pp. 237–271.

———, *Worlds in the Making.* New York and London: Harper & Brothers, 1908.

Bryson, R. A., and D. A. Baerreis, "Possibilities of Major Climate Modifi-

cation and Their Implications: Northwest India, a Case for Study,"
Bulletin of the American Meteorological Society 48 (1967), pp. 136–142.

Budyko, Mikhail I., "The Effect of Solar Radiation Variations on the Climate of the Earth," *Tellus* 21 (1969), pp. 611–619.

————, and Y. S. Sedunov, "Anthropogenic Climate Changes," *Proceedings of the World Congress,* Hamburg, November 1988.

Callendar, G. S., "On the Amount of Carbon Dioxide in the Atmosphere," *Tellus* 10 (1958), pp. 243–248.

Chamberlin, Thomas C., "An Attempt to Frame a Working Hypothesis of the Cause of Glacial Periods on an Atmospheric Basis," *Journal of Geology* 7 (1899), pp. 545–561.

The Conservation Foundation, *Implications of Rising Carbon Dioxide Content of the Atmosphere.* New York: The Conservation Foundation, 1963.

Greenhouse/Glasnost, Letter to President George Bush and Chairman Mikhail Gorbachev, after a meeting convened jointly by the Institute for Resource Management and the Soviet Academy of Sciences, Sundance, Utah, August 1989.

Jäger, Jill, and William W. Kellogg, "Anomalies in Temperature and Rainfall During Warm Arctic Seasons," *Climatic Change* 5 (1983), pp. 39–60.

Jones, P. D., T. M. L. Wigley, and P. B. Wright, "Global Temperature Variation Between 1861 and 1984," *Nature* 322 (1986), pp. 430–434.

Keeling, C. D., A. F. Carter, and W. G. Mook, "Seasonal, Latitudinal, and Secular Variations in the Abundance and Isotope Ratios of Atmospheric CO_2," *Journal of Geophysical Research* 89 (1984), pp. 4615–4628.

Kellogg, William W., "Mankind's Impact on Climate: The Evolution of an Awareness," *Climatic Change* 10 (1987), pp. 113–136.

————, and R. Schware, *Climate Change and Society: Consequences of Increasing Atmospheric Carbon Dioxide.* Boulder, Colo.: Westview, 1981.

————, and R. Schware, "Society, Science, and Climate Change," *Foreign Affairs* 60 (1982), pp. 1076–1109.

————, and Zong-ci Zhao, "Sensitivity of Soil Moisture to Doubling of Carbon Dioxide in Climate Model Experiments: Part I. North America," *Journal of Climate* 1 (1988), pp. 348–366.

MacKenzie, J. J., *Breathing Easier: Taking Action on Climate Change, Air Pollution, and Energy Insecurity.* Washington, DC: World Resources Institute, 1988.

Manabe, S., and R. T. Wetherald, "Thermal Equilibrium of the Atmosphere with a Given Distribution of Relative Humidity," *Journal of Atmospheric Science* 24 (1967), pp. 241–259.

Marland, G., "Fossil Fuels CO_2 Emissions: Three Countries Account for 50% in 1986," *CDIAC Communications* (Oak Ridge National Laboratory, TN), Winter 1989, pp. 1–3.

Meehl, G. A., and W. M. Washington, "A Comparison of Soil Moisture Sensitivity in Two Global Climate Models," *Journal of Atmospheric Science* 45, (1988), pp. 1476–1492.

Milankovitch, N., "Mathematische Klimalehre und astronomische Theorie der Klimaschwankungen," in W. Koppen and R. Geiger (eds.), *Handbuch der Klimatologie I(A).* Berlin: Gebrüder Borntraeger, 1930, p. 1.

Möller, F., "On the Influence of Changes in CO_2 Concentration in Air on the Radiation Balance of the Earth's Surface and the Climate," *Journal of Geophysical Research* 68 (1963), pp. 3877–3886.

PSAC, *Restoring the Quality of Our Environment: Report of the Environmental Pollution Panel.* Washington, DC: President's Science Advisory Committee, 1965.

Ralof, J., "Governments Warm to Greenhouse Action," *Science News* 136 (1989), pp. 394–397.

Ramanathan, V., "The Greenhouse Theory of Climate Change: A Test by an Inadvertent Global Experiment," *Science* 240 (1988), pp. 293–299.

————, H. B. Singh, R. J. Cicerone, and J. T. Kiehl, "Trace Gas Trends and Their Potential Role in Climate Change," *Journal of Geophysical Research* 90 (1985), pp. 5547–5566.

Reinhold, R., "How Fighting Global Warming Could Be Painless and Profitable," *The New York Times,* September 3, 1989.

Revelle, Roger, "Introduction: The Scientific History of Carbon Dioxide," in E. T. Sundquist and W. S. Broecker (eds.), *The Carbon Cycle and Atmospheric CO_2: Natural Variations Archean to Present* (Geophysical Monograph 32). Washington, DC: American Geophysical Union, 1985, pp. 1–4.

————, and Hans E. Suess, "Carbon Dioxide Exchange Between Atmosphere and Ocean and the Question of an Increase of Atmospheric CO_2 During the Past Decades," *Tellus* 9 (1957), pp. 18–27.

Rotty, R. M., and G. Marland, "Fossil Fuel Combustion: Recent Amounts,

Patterns, and Trends of CO_2," in J. R. Trabalka and D. E. Reichle (eds.), *The Changing Carbon Cycle: A Global Analysis.* New York: Springer-Verlag, 1986.

Schneider, Stephen H., "Doing Something About the Weather," *World Monitor* (*The Christian Science Monitor* monthly), December 1988, pp. 28–39.

————, *Global Warming: Are We Entering the Greenhouse Century?* San Francisco: Sierra Club, 1989.

Sellers, William D., "A Global Climate Model Based on the Energy Balance of the Earth-Atmosphere System," *Journal of Applied Meteorology* 8 (1969), pp. 329–400.

SMIC, *Inadvertent Climate Modification: Report of the Study of Man's Impact on Climate.* Cambridge, Mass.: MIT Press, 1971.

Tolman, C. F., Jr., "The Carbon Dioxide of the Ocean and Its Relations to the Carbon Dioxide of the Atmosphere," *Journal of Geology* 7 (1899), pp. 585–601.

Tyndall, John, "On Radiation Through the Earth's Atmosphere," *Philosophical Magazine* 4 (1863), pp. 200–207.

Villach/Bellagio, *Developing Policies for Responding to Climatic Changes.* Report (written by Jill Jäger) of workshops held in Villach, Austria (September 28–October 2, 1987), and Bellagio, Italy (November 9–13, 1987), sponsored by the World Meteorological Organization and the United Nations Environment Programme, WMO Report WCIP-1, WMO/TD No. 225, 1988.

Von Neumann, John, "Can We Survive Technology?" *Fortune,* June 1955.

Weinberg, A. M., "Nuclear Energy, Carbon Dioxide, and International Cooperation," in *Proceedings of the Conference on Technology-Based Confidence Building: Energy and Environment* (held at Santa Fe, NM, July 9–14, 1989). Berkeley: University of California, and Los Alamos, N. Mex.: Los Alamos National Laboratory, pp. 474–477.

Zhao, Zong-ci, and William W. Kellogg, "Sensitivity of Soil Moisture to Doubling of Carbon Dioxide in Climate Model Experiments: Part II. The Asian Monsoon Region," *Journal of Climate* 1 (1988), pp. 367–370.

STEPHEN H. SCHNEIDER

The Changing Climate: A Risky Planetwide Experiment

Stephen H. Schneider is currently head of the Interdisciplinary Climate Systems Section at the National Center for Atmospheric Research. He is the author of *Global Warming: Are We Entering the Greenhouse Century?*, *The Coevolution of Climate and Life,* and *The Genesis Strategy: Climate and Global Survival.* He has written or edited more than 160 scientific papers and is editor of the scientific journal *Climatic Change.*

A consultant to the Carter and Nixon administrations, Dr. Schneider is a frequent witness at congressional hearings on environmental and climatic issues. He was selected by *Science Digest* in 1984 as one of the "One Hundred Outstanding Young Scientists in America," and is a fellow of the American Association for the Advancement of Science. His current research focuses on global warming and climatic change, with emphasis on climatic modeling of paleoclimates and of human impact on climate (e.g., the greenhouse effect or nuclear war). Dr. Schneider is dedicated to advancing the public understanding of science.

Even though there is virtually no debate among scientists about the greenhouse effect as a scientific proposition, there is controversy. Will the rising concentrations of greenhouse gases raise the Earth's temperature by 1°, 5°, or 8°C? Will the increase take 50, 100, or 150 years? Will it be drier in Iowa, wetter in India? There is still more controversy when it comes to policy: Should steps be taken to reduce greenhouse warming or to anticipate its effects? What steps, and when? In the face of so much controversy, an understanding of what is known well, known slightly, and not at all known about the greenhouse warming is essential.

Circumstantial evidence from the geologic and historical past bears out a link between climatic change and fluctuations in greenhouse gases. Between 3.5 and 4 billion years ago the sun is thought to have been about 30 percent fainter than it is today. Yet life evolved and sedimentary rock formed under the faint young sun: the temperature of at least some of the earth's surface was above the freezing point of water. Some workers have proposed that the early atmosphere contained as much as 1,000 times today's level of carbon dioxide, which compensated for the sun's feeble radiation by its heat-trapping effect.

Later, an enhanced greenhouse effect may have been partly responsible for the warmth of the Mesozoic Era—the age of the dinosaurs—which fossil evidence suggests was perhaps 10° to 15°C warmer than today. At the time, 100 million years ago and more, the continents occupied different positions from those today, and thus the circulation of the oceans differed and the transport of heat from the tropics to high latitudes may have been greater. Yet calculations by Eric J. Barron, now at Pennsylvania State University, and others suggest that paleocontinental geography can explain no more than half of the Mesozoic warming.

Increased carbon dioxide can readily explain the extra heating, as Aleksandr B. Ronov and Mikhail I. Budyko of the Leningrad State Hydrological Institute first proposed and as Barron, Starley L. Thompson of the National Center for Atmospheric Research (NCAR), and I have calculated. A geochemical model constructed by Robert A. Berner and Antonio C. Lasaga of Yale University and the late Robert M. Garrels of the University of South

Florida suggests that the carbon dioxide may have been released by unusually heavy volcanic activity on the midocean ridges, where new ocean floor is created by upwelling magma.

Direct evidence linking greenhouse gases with the dramatic climatic changes of the ice ages comes from bubbles of air trapped in the Antarctic ice sheet by the ancient snowfalls that built up to form the ice. A team headed by Claude Lorius of the Laboratory of Glaciology and Geophysics of the Environment, near Grenoble, examined more than 2,000 meters of ice cores—a 160,000-year record—recovered by a Russian drilling project at the Vostok station in Antarctica. Laboratory analysis of the gases trapped in the core showed that carbon dioxide and methane levels in the ancient atmosphere correlated closely with each other and, more important, with the average local temperature (determined from the ratio between hydrogen isotopes in the water molecules of the ice).

During the current interglacial period (the past 10,000 years) and the previous one (a 10,000-year period around 130,000 years ago), the ice recorded a local (i.e. Antarctic) temperature about 10°C warmer than at the height of the ice ages. (The earth as a whole is about 5°C warmer during interglacials.) At the same time, the atmosphere contained about 25 percent more carbon dioxide and 100 percent more methane than during the glacial periods. It is not clear whether greenhouse gas variations caused the climatic changes or vice versa. My guess is that the ice ages were paced by other factors, such as changes in the earth's orbit and the dynamics of ice buildup and retreat, but biological changes and shifts in ocean circulation in turn affected the atmosphere's trace-gas content, amplifying the climatic swings.

A still more detailed record of greenhouse gases and climate comes from the past 100 years, which have seen a further 25 percent increase in carbon dioxide above the interglacial level and another doubling of atmospheric methane. Two groups, one led by James E. Hansen at the National Aeronautics and Space Administration's Goddard Institute for Space Studies and the other by T. M. L. Wigley at the Climatic Research Unit of the University of East Anglia, have constructed records of global average surface temperature for the past century. The workers drew on data from many of the same recording stations around the globe (the Climatic Research Unit also included readings made at sea), but they had different techniques for analyzing the records and compensating for their shortcomings. Certain recording stations were moved over the course of the century, for example, and readings from city centers may have been skewed by heat released by machinery or stored by buildings and pavement.

This "urban heat island" effect is likely to have been disproportionately large in developed countries such as the United States, but even when the same correction calculated for the U.S. data (by Thomas R. Karl of the National Climatic Data Center in Asheville, North Carolina, and P. D. Jones of East Anglia) is applied to the global data set, about 0.5°C of unexplained "real" warming over the past 100 years remains in both records. In keeping with the trend, the 1980s appear to be the warmest decade on record, and 1988, 1987, and 1981 the warmest years, in that order. These warm years were later reconfirmed by NASA satellite measurements of the temperature of the atmosphere for the 1980s.

Is this record warm decade the signal of the greenhouse warming? It is tempting to accept it as such, but the evidence is not definitive. For one thing, instead of the steady warming one might expect from a steady buildup of greenhouse gases, the record shows rapid warming until the end of World War II, a slight cooling through the mid–1970s, and a second period of rapid warming since then.

What trajectory will the temperature curve follow now? Three basic questions must be answered in forecasts of the climatic future: How much carbon dioxide and other greenhouse gases will be emitted? By how much will atmospheric levels of the gases increase in response to the emissions? What climatic effects will the resulting buildups have, after natural and human factors that might mitigate or amplify those effects are taken into account?

Projecting emissions is an intricate exercise in social science. How much carbon dioxide humanity as a whole will be emitting in the future depends primarily on the global consumption of fossil fuels and the rate of deforestation (which accounts for perhaps half of the buildup since the year 1800, and 20 percent of current emissions). Each factor in turn is affected by many others. Increased use of fossil fuels, for example, will reflect population growth, the rate at which alternative energy sources and conservation measures are adopted, and the state of the world economy. Typical projections assume that global fossil-fuel consumption will continue increasing at about its current pace—much slower than it grew before the energy crisis of the 1970s—yielding increases in carbon dioxide emissions of between 0.5 and 2 percent a year for the next several decades at least.

Other greenhouse gases, such as methane, CFCs, nitrogen oxides, and low-level ozone, together could contribute as much to global warming as carbon dioxide, even though they are emitted in much smaller quantities: they are much better at absorbing infrared radiation. But predicting future emissions for these gases is even more complicated than it is for carbon dioxide. The sources of some gases, such as methane, are not well understood; the production of other gases, such as CFCs and low-level ozone, could rise or fall sharply depending on whether specific technological or policy steps are taken.

If we assume a plausible scenario for future carbon dioxide emissions, how fast will the atmospheric concentration increase in response? Atmospheric carbon dioxide is continuously being absorbed by green plants and in chemical and biological processes in the oceans. The rate of carbon dioxide uptake is likely to change as the atmospheric concentration changes; that is, feedback processes will enter the equation. Because carbon dioxide is a raw material of photosynthesis, an increased concentration might speed the uptake by plants, counteracting some of the buildup. Similarly, because the carbon dioxide content of the oceans' surface waters stays roughly in equilibrium with that of the atmosphere, oceanic uptake will slow the buildup to some extent. (The slower the buildup is in the first place, the more effective, proportionally, oceanic uptake is likely to be.)

It is also possible, however, that an increased concentration of carbon dioxide and other greenhouse gases will trigger positive feedbacks that would add to the atmospheric burden. Rapid change in climate could disrupt forests and other ecosystems, reducing their ability to draw carbon dioxide down from the atmosphere. Moreover, climatic warming could lead to rapid release of the vast amount of carbon held in soil as dead organic matter. This stock of carbon—at least twice as much as is stored in the forests or the atmosphere—is continuously being decomposed into carbon dioxide and methane by the action of soil microbes. A warmer climate might speed their work and release additional carbon dioxide (from dry soils) and methane (from rice paddies, landfills, and wetlands), which would enhance the warming. Large quantities of methane are locked up also in continental-shelf sediments and below Arctic permafrost in the form of clathrates, molecular lattices of methane and water. Warming of the shallow waters of the oceans and melting of the permafrost could release some of the methane.

In spite of these uncertainties, many expect uptake by plants and by the oceans to moderate the carbon dioxide buildup, at least for the next 50 or 100 years. Typical estimates, based on current or slightly increased emission rates, put the fraction of newly injected carbon dioxide that will remain in the atmosphere at about one-half. Under that assumption, the atmospheric concentration will reach 600 parts per million, or about twice the level of 1900, by sometime between 2030 and 2080. Some other greenhouse gases are expected to build up faster than carbon dioxide, however.

What effect will a doubling of atmospheric carbon dioxide have on climate? The historical record offers no clear quantitative guidance. Nor can climate—the product of complicated interactions involving the atmosphere, the oceans, the land surface, vegetation, and polar ice—be physically reproduced in a laboratory experiment. In exploring the future of the earth's climate, my colleagues and I rely on mathematical climate models called global circulation models (GCMs).

The models, which have been built at Princeton University's Geophysical Fluid Dynamics Laboratory, the Goddard Institute for Space Studies, here at NCAR, and elsewhere, consist of expressions for the interacting components of the ocean-atmosphere system and equations representing the basic physical laws governing their behavior, such as the ideal gas laws and the conservation of mass, momentum, and energy. Given values for, say, the input of energy from the sun and the composition of the atmosphere, a model calculates "climate"—temperature and, in sophisticated models, pressure, wind speed, humidity, soil moisture, and other variables.

To keep the task computationally manageable, the calculations are done at discrete points in a simplified version of the real world. In spite of the simplification, running such a GCM for only one simulated year can take many hours on the fastest available supercomputers.

To study the effect of a trace-gas buildup, a modeler simply specifies the projected amount of greenhouse gases and compares the model results with a control simulation of the existing climate, based on the present atmospheric composition. The results of the most recent GCMs are in rough agreement: A doubling of carbon dioxide, or an equivalent increase in other trace gases, would warm the Earth's average surface temperature by between 2° and 5°C. Such a change would be unprecedented in human history. At the high end of the range, it would match the 5°C warming since the peak of the last ice age 18,000 years ago but would take effect some 100 times faster.

The shortcomings of computer models limit the reliability of such forecasts. Many processes that affect global climate are simply too small to be seen at the coarse resolution of a model. Such climatically important processes as atmospheric turbulence, precipitation, and cloud formation take place on a scale not of hundreds of kilometers (the scale of the grid in a GCM) but of a few kilometers or less. Since such processes cannot be simulated directly, modelers must find a way of relating them to variables that can be simulated on the model's coarse scale. They do so by developing a parameter—a proportionality coefficient—that relates, say, the average cloudiness within a grid cell to the average humidity and temperature (something the model can calculate).

This strategy, known as parameterization, has the effect of aggregating small-scale phenomena that could act as feedbacks on climatic change, either amplifying or moderating it. Clouds, for example, reflect sunlight back to outer space (and thus tend to cool the climate) and also absorb infrared radiation from the earth (and thus warm it). Which effect dominates depends on the clouds' brightness, height, distribution, and extent. Recent satellite measurements have confirmed two-decade-old calculations showing that clouds currently have a net cooling effect; the earth as a whole would be much warmer under cloudless skies. But climatic change might cause incremental changes in cloud characteristics, altering the nature and amount of the feedback. Present models, crudely reproducing only average cloudiness, can say little that is reliable about cloud feedback—or about the many other feedbacks that depend on parameterized processes.

Another shortcoming of present models is their crude treatment of the oceans. The oceans exert potent effects on the present climate and will surely influence climates to come. Their enormous thermal mass will act as a "thermal sponge," slowing any initial increase in global temperature while the oceans themselves warm up. The magnitude of the effect will depend on ocean circulation, which in turn may change as the earth warms. In principle, a climate model should couple a simulated atmosphere with oceans whose dynamics are simulated in comparable detail. The computational challenge is staggering, however, and in most GCMs applied to greenhouse warming the dynamics of the oceans are simplified, treated at coarse resolution, or left out.

In addition to limiting the reliability of global forecasts, the simplified treatment of the oceans also prevents the models from giving a definitive picture of how climate will change over time in specific regions. Ideally one would like to know not only how much the world as a whole will warm but also whether it will, say, get drier in Iowa, wetter in India, or more humid in New York City. Yet as long as the oceans are out of equilibrium with the atmosphere, their thermal effects will be felt differently in different places. An area in which there is little mixing between surface waters and cold, deep waters might warm quickly; high-latitude regions where deep water is mixed even to the surface might warm more slowly. These thermal effects could in turn affect wind patterns, thereby altering other regional variables, including humidity and rainfall. (Regional forecasts are also compromised in many models by simplified representations of vegetation, which ignore climatically important processes such as the release of water vapor by plants and their effect on surface albedo, or reflectiveness.)

Nevertheless, climatologists have grounds for considerable confidence in their models' forecasts of global surface-temperature change. Individual model elements can be verified by comparing them with the results of a more detailed submodel—a smaller, finer-scale simulation—or with real data. Cloud parameterizations, for example, can be tested against actual measurements of the relation of temperature and humidity to cloudiness within an area corresponding to a cell in the model.

The skill of a model as a whole, and in particular its ability to account for relatively fast processes, such as changes in atmospheric circulation or average cloudiness, can be verified by checking its ability to reproduce the seasonal cycle—a twice-yearly change in hemispheric climate that is larger than any projected greenhouse warming. In spite of parameterization, most GCMs map the seasonal cycle of surface temperature quite well, but their ability to simulate seasonal changes in other climatic variables, including precipitation and relative humidity, has not been studied as thoroughly.

During the course of decades (the expected time scale for unmistakable global warming), other, slower processes that do not affect the seasonal cycle come into play: changes in ocean currents or in the extent of glaciers, for instance. Simulations of past climates—the ice ages or the Mesozoic hothouse—serve as a good check on the long-term accuracy of climate models. To such tests of overall validity can be added simulations of the climates of other planets, such as Venus, where a dense greenhouse atmosphere maintains a surface temperature of about 450°C.

The record of the past hundred years provides the only direct test of the models' ability to simulate the effects of the ongoing greenhouse gas increase. When a climate model is run for an atmosphere with the composition of a century ago, and then run again for the historical 25 percent increase in carbon dioxide and doubling in methane, does it "predict" the observed 0.5°C warming? Actually, most models yield a somewhat larger warming, of at least a degree.

If the observed temperature increase really is a sign of greenhouse warming and not just "noise"—a random fluctuation—one might account for the disparity in various ways. Perhaps the models are simply twice too sensitive to small increases in greenhouse gases, or perhaps the incomplete and inhomogeneous network of thermometers has underestimated the global warming. It is conceivable that some other factor, not well accounted for in the models, is delaying or counteracting the warming. It might be that the heat capacity of the oceans is larger than current models calculate, that the sun's output has declined slightly, or that volcanoes have injected more dust into the stratosphere than is currently known, thereby reducing the solar energy reaching the ground.

It may be significant that the transient cooling interrupting the warming trend began around 1940 and was most pronounced in the Northern Hemisphere, coinciding in time and place with a sharp increase in emissions of sulphur from coal- and oil-burning factories and power plants. The sulphur, a major cause of acid rain, is emitted as a gas, sulphur dioxide, but is transformed into fine sulphate particles once in the atmosphere. The particles can travel long distances and serve as condensation nuclei for the formation of cloud droplets, and so they may make some clouds brighter, and thus increase their cooling effects. In addition, if no soot is bound to the sulphate, it forms a reflective haze even in cloudless skies. Sulphur emissions could be one factor that has held a greenhouse warming down somewhat in the Northern Hemisphere, especially since World War II.

The discrepancy between the predicted warming and what has been seen so far keeps most climatologists from saying with great certainty (99-percent confidence, say) that the greenhouse warming has already taken hold. Yet the discrepancy is small enough, the models are well enough validated, and other evidence of greenhouse gas effects on climate is strong enough that most of us believe the increases in average surface temperature predicted by the models for the next fifty years or so are probably valid within a rough factor of 2. (By "probably" I mean it is a better than even bet.) Within a decade or so, warming of the predicted magnitude should be clearly evident, even in the noisy global temperature record. But waiting for such conclusive, direct evidence is not a cost-free proposition: by then the world will already be committed to greater climatic change than it would be if action were taken now to slow the buildup of greenhouse gases. Of course, whether or not to act is a value judgment, not a scientific matter.

Why worry about changes in climate on the scale predicted by the models? Changes in temperature and precipitation could threaten natural ecosystems, agricultural production, and human settlement patterns. Particular forest types, for example, grow in geographic zones defined largely by temperature. The belt of spruce and fir that now spans Canada grew far to the south at the end of the last ice age 10,000 years ago, hugging the edge of the ice sheet. As the climate warmed at an average rate of one or two degrees every 1,000 years and the ice retreated, the forest belt migrated northward, at perhaps a kilometer a year. Forests probably could not sustain the much faster migration required by the projected warming, and many ecosystems cannot migrate in any case: they exist only in preserves, which might become marooned in a newly inhospitable climate zone.

Human activities could be affected directly if a warming speeded the evaporation of moisture, reducing stream runoff; in the western United States a temperature increase of several degrees could decrease runoff in the Colorado basin substantially even if precipitation held steady. As water ran short, faster evaporation would increase the demand for irrigation and add to the strain on water supplies. At the same time, water quality might suffer as the same waste volume was concentrated in lower stream volumes.

What is more, several climate models predict that summer precipitation will actually decline in midcontinental areas, including the central plains of the United States. The late Dean F. Peterson, Jr., of Utah State University, and Andrew A. Keller of Keller-Bliesner Engineering in Logan, Utah, estimated the effects on crop production of a 3°C warming combined with a 10-percent drop in precipitation. They found that with increased crop water needs and a reduction in available water, the viable acreage in arid regions of the western states and the Great Plains would fall by nearly a third. (A western drying might also result in an increased frequency of wildfires.)

Coastal areas, meanwhile, might face a rise in sea level. Most workers expect a global temperature increase of a few degrees over the next 50 or 100 years to raise sea level by between 0.2 and 1.2 meters as a result of the thermal expansion of the oceans, the melting of mountain glaciers, and the possible retreat of the Greenland ice sheet's southern margins. (Ice could actually build up in Antarctica because of warmer winters, which would probably increase snowfall.) The rising sea would endanger coastal settlements and ecosystems and might contaminate groundwater supplies with salt. In spite of many local factors that make it difficult to isolate a consistent global signal, one group of researchers recently claimed to have found a uniform worldwide rise in sea level of about 2 millimeters a year in long-term tide-gauge records. That rise is somewhat larger, however, than one would have expected from the warming seen so far.

Clearly, these direct effects of climatic change would have powerful economic, social, and political consequences. A decline in agricultural productivity in the Midwest and Great Plains, for example, could be disastrous for farmers and the U.S. economy. Cutting into the U.S. grain surplus might have serious implications for international security.

To be sure, not everyone would lose. If the corn belt simply moved north by several hundred kilometers, for example, Iowa's billion-dollar loss could become Minnesota's billion-dollar gain. But how could the losers be compensated and the winners charged? The issue of equity would become still more thorny if it spanned borders—if the release of greenhouse gases by the economic activities of one country or group of countries did disproportionate harm to other countries whose activities had contributed less to the buildup.

In the face of this array of threats, three kinds of responses could be considered. First, some have proposed technical measures to counteract climatic change—deliberately spreading dust in the upper atmosphere to reflect sunlight, or fertilizing oceans with iron to stimulate growth of plankton that remove CO_2, for instance. Yet if unplanned climatic changes themselves cannot be predicted with certainty, the effects of such countermeasures would be comparably unpredictable. Such "technical fixes" would run a real risk of misfiring—or of being blamed for any unfavorable climatic fluctuations that took place at the same time.

Many economists tend to favor a second class of action: adaptation, often with little or no attempt to anticipate damages or prevent climatic change. Adaptive strategists argue that the large uncertainties in climate projections make it unwise to spend large sums trying to avert outcomes that may never materialize. They argue that adaptation, in contrast, is cheap: the infrastructure that would have to be modified in the face of climatic change—such as water-supply systems and coastal structures—will have to be replaced in any case before large climatic changes are due to appear. The infrastructure can simply be rebuilt as needed to cope with the changing environment.

Passive adaptation relies mostly on reacting to events as they unfold, but some active adaptive steps could be taken now to make future accommodation easier. An American Association for the Advancement of Science panel on climatic change made a strong, potentially controversial but, I believe, compelling suggestion for active adaptation: governments at all levels should reexamine the technical features of water systems and the economic and legal aspects of water-supply management in order to increase the systems' efficiency and flexibility. As the climate warms and precipitation and runoff change, water shortages may grow more common and needs for regional transfers more complex. Even if climate did not change, more flexible water systems would make it easier to cope with the normal extremes of weather.

The third and most active category of response is prevention: curtailing the greenhouse-gas buildup. Energy-conservation measures, alternative energy sources or a switch from coal to natural gas and other fuels with a lower carbon content could all reduce carbon dioxide emissions, as could a halt to deforestation. Stopping the production of CFCs, already notorious because of their ability to erode the stratospheric ozone layer, would eliminate another component of the buildup. A far-reaching proposal for an international framework for reducing emissions was put forward in 1976 by Margaret Mead and William W. Kellogg of NCAR: a "law of the air," which would keep emissions of carbon dioxide below a global standard by assigning polluting rights to each nation.

Proposals for immediate action are controversial because they often entail large immediate investments as insurance against future events whose details are far from certain. Is there some simple principle that can help us to choose which preventive or adaptive measures to spend our resources on? I believe it makes sense to take actions that will yield "tie-in" benefits even if climatic changes do not materialize as forecast.

Pursuing energy efficiency is a good example of a tie-in strategy. More efficient fossil-fuel use will slow the carbon dioxide buildup, and even if the sensitivity of climate to carbon dioxide has been overstated, what would be wasted by taking this step? Efficiency usually makes economic sense, and a reduction in fossil-fuel use would curb acid rain and urban air pollution and lessen the dependence of many countries on foreign producers. Developing alternative energy sources, revising water laws to add flexibility, searching for drought-resistant crop strains, negotiating international agreements on trade in food and other climate-sensitive goods—these steps could also offer widespread benefits even in the absence of any climatic change.

Often such steps will nonetheless be costly and politically controversial. Regulations or incentives to foster energy-efficient technologies might burden some groups—coal miners and the poor, perhaps—more than others, and the costs may be proportionally greater for poor countries than for rich ones. Actions to prevent a greenhouse warming will have to be coupled with domestic- and foreign-policy measures that attempt to balance fairness and effectiveness. Still, I believe it is better to fight poverty and foster development through direct investment rather than through artificially low energy prices that neglect the costs of the resulting environmental disruptions.

Some people argue that the free market, not government regulation or tax incentives, should dictate increases in energy efficiency, say, or the elimination of CFCs. But it cannot be logically argued that the market is "free" when it does not include some of the potential costs of environmental damage caused by goods or services. Moreover, even political conservatives agree that an economic calculus must give way to a strategic consciousness when national or global security is at stake.

Security is indeed at stake here, as the implications of a global temperature rise of several degrees or more over the next century make clear.

Adding to the predicted threats are surprises that may be lurking in the greenhouse century: a sharp positive feedback in the greenhouse gas buildup from accelerated decay of soil organic matter, dramatic changes in regional climates because of a shift in ocean circulation, or the outbreak of new diseases or agricultural pests as ecosystems are disrupted. In my value system—and this is a political and not a scientific judgment—effective tie-in actions are long overdue.

I am often asked whether I am pessimistic because it will be impossible to avert some global change: at this stage, it appears, no plausible policies are likely to prevent the world from warming by a degree or two. Actually, I see a positive aspect: the possibility that a slight but manifest global warming, coupled with the larger threat forecast in computer models, may catalyze international cooperation to achieve environmentally sustainable development, marked by a stabilized population and the proliferation of energy-efficient and environmentally safe technologies. A much larger greenhouse warming (together with many other environmental disruptions) might thereby be averted.

The developed world might have to invest hundreds of billions of dollars every year for many decades, both at home and in financial and technical assistance to developing nations, to achieve a stabilized and sustainable world. It is easy to be pessimistic about the prospects for an international initiative of this scale, but not long ago a massive disengagement of NATO and Warsaw Pact forces in Europe also seemed inconceivable. Disengagement now seems possible, even likely, to me. Perhaps the resources such an agreement would free and the model of international cooperation it would provide could open the way to a world in which the greenhouse century exists only in the microchips of a supercomputer.

REFERENCES

Bolin, Bert, B. R. Döös, Jill Jäger, and Richard A. Warrick, *The Greenhouse Effect, Climatic Change, and Ecosystems.* New York: Wiley, 1986.

Intergovernmental Panel on Climate Change, *Working Group I* (second draft report). Geneva: World Meteorological Organization, March 1990.

Ramanathan, V., et al., "Cloud-Radiative Forcing and Climate: Results from the Earth Radiation Budget Experiment," *Science* 243, no. 4887 (January 6, 1989), pp. 57–63.

Schneider, Stephen H., *Global Warming: Are We Entering the Greenhouse Century?* San Francisco: Sierra Club, 1989.

Washington, Warren M., and Claire L. Parkinson, *An Introduction to Three-Dimensional Climate Modeling.* Mill Valley, CA: University Science, 1986.
Wigley, T. M. L., "Possible Climate Change Due to SO_2-Derived Cloud Condensation Nuclei," *Nature* 339, no. 6223 (June 1, 1989), pp. 365–367.

JAMES E. LOVELOCK

The Evolution of Gaia

James Lovelock is president of the Marine Biology Association, an independent scientific researcher, and the inventor of many instruments used in chemical analysis. One of these devices, the electron capture detector, has been crucial to the development of the environmental movement: it revealed, for the first time, the ubiquitous presence of pesticide residues and has confirmed the global distribution of nitrogen oxides and chlorofluorocarbons. Several of Dr. Lovelock's inventions have been adopted by NASA for use in planetary exploration programs. He was awarded three certificates of recognition by NASA for his contributions.

Dr. Lovelock has taught at Baylor College of Medicine, the University of Houston, and the University of Reading. He is the author of two books, *Gaia: A New Look at Life on Earth* (Oxford University Press, 1979) and *The Ages of Gaia* (Norton, 1988), and of many scientific papers and articles. He has received numerous awards, including the American Chemical Society's Award for Chromatography, the Silver Medal and Prize of the Plymouth Marine Laboratory, and the Norbert Gerbier Prize of the World Meteorological Organization. Dr. Lovelock was made a Commander of the British Empire in 1990. He was just awarded the Amsterdam Prize for the Environment by the Royal Netherlands Academy of Arts and Sciences.

BEGINNINGS

The Gaia hypothesis is now twenty years old. Those who have worked closely with it deem the hypothesis reasonable science and see it growing ever more plausible as evidence and models map together. But theirs is a minority opinion. Most scientists regard Gaia as little more than metaphor; some even denounce it as antiscience. This commentary is about the history, arguments, evidence, and criticisms of Gaia, and about where it now stands.

Gaia began in the mid–1960s at the Jet Propulsion Laboratory in California. The National Aeronautics and Space Administration had sought my help in its quest to discover whether there was life on Mars. For the most part the experiments proposed by NASA biologists were geocentric, seeking to find on Mars the life they were familiar with in their laboratories here. Better, I suppose, to do this than to seek a speculative form of life, but surely there was a better and more certain way to detect Martian life. In a letter to *Nature* in 1965, I proposed some physical tests for the presence of planetary life.[1] One of these was a top-down view of the entire planet instead of a local search at the site of landing. The test was simply to analyze the chemical composition of the planet's atmosphere. If the planet were lifeless, then it would be expected to have an atmosphere determined by physics and chemistry alone, and to be close to the chemical equilibrium state. But if the planet bore life, organisms at the surface would be obliged to use the atmosphere as a source of raw materials and as a depositary for wastes. Such a use of the atmosphere would change its chemical composition. It would depart from equilibrium in a way that would show the presence of life.

Together with Dian Hitchcock I examined atmospheric evidence from the infrared astronomy of Mars.[2] We compared this evidence with that available on the sources and sinks of atmospheric gases of the one planet we knew bore life, Earth. We found an astonishing difference. The atmosphere of

1. J. E. Lovelock, *Nature* 207 (1965), pp. 568–570.
2. D. R. Hitchcock and J. E. Lovelock, *Icarus* 7 (1967), pp. 149–159.

Mars was close to chemical equilibrium and dominated by carbon dioxide, but that of the Earth was in a state of deep chemical disequilibrium. In our atmosphere, carbon dioxide is a mere trace gas. The coexistence of abundant oxygen with methane and other reactive gases would be impossible on a lifeless planet. Even Earth's abundant nitrogen and water are difficult to explain by geochemistry. There are no such anomalies in the atmospheres of Mars and Venus, and their existence in the Earth's atmosphere signals the presence of living organisms at the surface. Sadly, we concluded, Mars is probably lifeless.

That first sight from space of a white-and-blue-dappled sphere irreversibly altered our emotions about the Earth. Atmospheric chemistry seen from above was, in scientific terms, for me a similar revelation of the Earth. This top-down analysis revealed the atmosphere as a gas mixture like that of the intake manifold of an internal combustion engine, with oxygen and combustible gases mixed, and different from the exhausted, carbon dioxide–dominated atmospheres of Mars and Venus. Further, I knew that the chemical composition of the atmosphere was stable for long periods when compared with the residence times of its gases. As I thought about these facts one afternoon in 1965 at the Jet Propulsion Lab, I had a flash of enlightenment: Such atmospheric constancy required the existence of an active control system.

Then, I lacked any idea of the nature of this control system, except that the organisms on the Earth's surface were part of it. I learned from astrophysicists that stars increase their heat output as they age and that our sun has grown in luminosity by 25 percent since life began. I realized that, in the long term, climate also might be actively regulated. The notion of a control system involving the whole planet and the life on it was now firmly established in my mind. Sometime near the end of the 1960s I discussed this idea with a neighbor, the novelist William Golding. He suggested "Gaia"—the name the ancient Greeks gave to personify the Earth—as the only one appropriate for so powerful an entity. I shall use the term "Gaia" as shorthand for the theory.

I first stated the idea of Gaia as a control system in 1972.[3] Shortly afterwards I began a continuing collaboration with Lynn Margulis. We first stated our hypothesis in words such as these: "Life, or the biosphere, regulates or maintains the climate and the atmospheric composition at an optimum for itself." By 1973 we refined the hypothesis and restated Gaia in papers in *Tellus*[4] and *Icarus*[5]: "The notion of the biosphere as an

3. J. E. Lovelock, *Atmospheric Environment* 6 (1972), pp. 579–580.
4. J. E. Lovelock and L. Margulis, *Tellus* 26 (1974), pp. 1–10.
5. L. Margulis and J. E. Lovelock, *Icarus* 21 (1974), pp. 471–489.

adaptive control system that can maintain the Earth in homeostasis, we are calling the Gaia hypothesis."

TELEOLOGY

Neither I nor Lynn Margulis has ever proposed a teleological hypothesis. Nowhere in our writings do we express the idea that planetary self-regulation is purposeful or involves foresight or planning by biota. It is true our early statements about Gaia were imprecise and open to misinterpretation, but this does not justify the persistent, almost dogmatic, criticism that our hypothesis is teleological. Sometimes the criticism reveals an ignorance of the circular logic of control theory. How often are the terms "negative feedback" and "positive feedback" used, through lack of understanding, as mere metaphors?

As an inventor I know it is easy to imagine, construct, and then reduce a self-regulating device to practice. I also know how difficult it is to explain it in words or even in analytical mathematics. Like an invention, Gaia is in many ways difficult to describe. The nearest I can come is to call Gaia the theory of an evolving system—a system made from the living organisms of the Earth, and from their material environment, the two parts tightly united and indivisible. This evolutionary theory views the self-regulation of climate and chemical composition as emergent properties of the system. The emergence is entirely automatic; no teleology is invoked. Gaia evolves as a system gradually, during long periods of homeostasis punctuated by sudden simultaneous changes in both organisms and environment. These changes move the system to new and different homeostatic states; a significant jump of this kind occurred between the anaerobic Archean and the oxygenated Proterozoic 2.5 thousand million years ago. Gaia offers a resolution of the long debate over whether evolution was gradual or punctuated, by suggesting it was both.

This is not a new theory of evolution. James Hutton intuited it when, in 1788, he saw the Earth as a superorganism whose proper study was by physiology. Alfred Lotka expressed it in 1925 when he stated clearly that the evolution of organisms could not be separated from the evolution of their physical environment. Gaia is not contrary to Charles Darwin's great vision; it includes the evolution of organisms by natural selection as an essential part of the planet's self-regulation.

Much of the confusion about Gaia and Darwinism comes from the misinterpretation of the concept of adaptation. My colleague Andrew Watson succinctly expressed the step that distinguishes Gaia from Darwinism in a debate before the Linnaean Society in December 1989. It lies in the tightness of the coupling between organisms and their physical environment. Almost everyone, Watson observed, now accepts that life influences the environment profoundly; this is now the conventional wisdom among geochemists, and a considerable change from their view before Gaia. It is equally obvious, Watson continued, that life is influenced by and adapts to the environment. This is the older wisdom that has prevailed throughout this century. Therefore, Watson concluded, life and the environment are a coupled feedback system; changes in one element will affect the other, and this influence may in turn feed back on the original change. The real debate, then, is: How important and how tight is the coupling? Does it, as we believe, confer new properties on the system, such as enhanced stability or behavior like that of a living organism?

HOLISM

Gaia asserts that the close coupling of organisms and their environment is strong enough to have influenced greatly the way in which the life-environment system on Earth, and on other planets with life, has evolved. It is strong enough that we will not properly understand Earth history until we think of the system as just that, a whole system, and stop trying to understand its parts in isolation from one another.

Like living organisms generally, Gaia is an open but bounded system. On one side the bound is space, and Gaia exchanges radiation across it. The other bound is the base of the crustal rocks, where matter exchanges with the near-infinite mass of Earth's mantle. Gaia's wastes are infrared radiation that escapes to space, and rock subducting to merge with the mantle. The profound disequilibrium of the atmosphere is one measure of the entropy reduction of this system. I think James Hutton was right to call the Earth a superorganism and to regard physiology as the proper science for its investigation.

In 1982, the biologist Richard Dawkins made the strong criticism that there is no way for evolution by natural selection to lead to altruism on a global scale.[6] My answer was the numerical model Daisyworld. The model is of a simple planet that keeps its climate constant in the face of ever-increasing solar output. It does so by the competition for space of two species of daisies, one dark and one light. The planetary albedo (the fraction

6. R. Dawkins, *The Extended Phenotype* (New York: Oxford, 1982).

of incident light or electromagnetic radiation that is reflected by a surface or body such as clouds or the moon) is tightly coupled to the evolution of the daisies, and the evolution of the daisies is tightly coupled to the evolution of the climate.

Daisyworld is a robust model and the key to understanding Gaia.[7] It employs realistic planetary and daisy properties and will withstand a wide range of choices in establishing its initial conditions. Daisyworld, like nature, is defined by nonlinear equations; no mathematical fudges, such as linear abstractions, are required. Most important, the Daisyworld model illustrates how Gaian regulation evolves. Always starting from the action of an individual organism, Daisyworld comes alive when the heat from its star is strong enough to allow growth and germination. Dark daisies absorb more sunlight than light-colored ones. When the star is cool, the dark daisies stay warmer; they can grow and leave seeds. The proliferation of dark daisies affects first themselves, then their local environment, and lastly the planet. When the whole planet is warm, competition between dark and light daisies ensures a stable climate despite ever-increasing heat from the star. Eventually, the system ends in catastrophe when the cooling effect of a planetwide growth of light-colored daisies is not enough to stabilize the planet's temperature. Daisyworld illustrates control theory: The *attainment* of stability depends on strong positive feedback; the *maintenance* of stability depends on negative feedback, and on the correct timing of its application. Understanding this is important; even geophysicists use the terms negative and positive feedback too loosely—as if one were always stabilizing and the other destabilizing.

Daisyworld has evolved. Instead of a make-believe world of daisies, there are now more sophisticated models that illustrate the simultaneous self-regulation of climate and chemistry by bacterial ecosystems. These new Gaia models provide a plausible account of the evolution of the Earth in the Archean and Proterozoic periods.[8]

Gaia models are general postulations, not limited to a specific set of criteria of interest, as are models drawn from the separated disciplines of science. The Gaia models can include many species or ecosystems simultaneously and follow their regulation as well as that of the material environment. Population biologists, for sixty years, have been unable to

<hr>

7. J. E. Lovelock, "Gaia as seen through the atmosphere," in *Biomineralisation and Biological Metal Accumulation* (ed. P. Westbroek and E. W. deJong), (Dordrecht: D. Reidel Publishing Company, 1983), pp. 15–25.

8. J. E. Lovelock, *The Ages of Gaia* (New York: Norton, 1988); J. E. Lovelock, *Revised Geophysics* 27, no. 2 (1989), pp. 215–222.

model more than two species simultaneously. The box models of biogeo-
chemists also are unstable and unduly sensitive to the choice of initial
conditions. These modelers handicap themselves by failing to consider
biota, or the material environment, interactively. Attempts to model the
world from the science of a single discipline seem doomed to fail, often by
intractable mathematical instability.

Some geochemists have been opponents of Gaia from the beginning.[9]
Their criticism is the reasonable one that Gaia is not needed to explain the
facts of the Earth. Geochemistry and geophysics, or even biogeochemistry,
they say, can do it alone. Yet these scientists are ready to tolerate Gaia as
a theory under test and find it interesting as a source of ideas for new
experiments. In an act of courage and generosity, climatologist Stephen
Schneider organized a conference of the American Geophysical Union in
1988 solely for discussions on Gaia. The conference is described in
Schneider's book *Global Warming*.[10] This meeting marked the end of the
false accusation of teleology—what some, illogically, have called the strong
Gaia hypothesis. At the conference, J. W. Kirchner made a spirited attempt
to demolish all notions of Gaia. Like some figure of the Inquisition, he
publicly burned several imaginary Gaias, and his pyrotechnic demolition of
the strong Gaia stole the show. But when the sparks faded, the real system
Gaia remained, hidden only by the smoke. The flux of papers inspired by
Gaia, now appearing in the journals, are the real proof of the conference's
value. Kirchner's actions have not stopped peer review from censoring any
mention of Gaia by name.

The value of Gaia as a theory is illustrated by the interpretation of the
long-term history of climate it makes possible. The conventional wisdom
about the problem of a cool early sun is that sufficient warmth for the start
of life, some 3.6 to 4 billion years ago, came from a greenhouse blanket
of atmospheric carbon dioxide. That would explain the start, but there is
no satisfactory explanation for the constancy of Earth's climate throughout
history—how temperatures have remained comfortable in spite of an ever-
increasing solar luminosity. Until the industrial revolution, carbon dioxide
has declined in abundance since the start of life by a factor of one thousand.
But how? No explanation exists for how the Earth has stayed warm. Early
photosynthetic life, by direct consumption and by increasing the rate of
weathering, must have removed carbon dioxide from the air, eating the

9. H. D. Holland, *The Chemical Evolution of the Atmosphere and the Oceans* (Princeton: Princeton
University, 1984); J. C. G. Walker, P. B. Hays, and J. Kasting, *Journal of Geophysical Research* 86
(1982), pp. 9776–9782.

10. S. H. Schneider, *Global Warming* (San Francisco: Sierra Club, 1989).

blanket that kept it warm. With a cool sun, this could have threatened a planetary-scale glaciation.

GEOPHYSIOLOGY

Geophysiology—my own neologism, but a useful one—offers a plausible explanation to the dilemma of warmth in the Earth's climatic history. In a model of the Archean period, a symbiotic pair of ecosystems, photosynthesizers and anaerobes, evolve tightly coupled with their physical and chemical environment.[11] The model system sustains a climate and atmospheric composition comfortable for Archean life. The photosynthesizers tend to cool by removing CO_2, both directly and by increasing the rate of weathering. The anaerobes tend to warm by producing the greenhouse gas methane. Organisms are assumed to grow best within the temperature range zero to fifty degrees Celsius. The model predicts an Archean atmosphere with methane near 0.1 percent, the chemically dominant gas, carbon dioxide, slightly less than 1 percent, and only traces of oxygen. The predicted global mean temperature is close to 20°C. At the end of the Archean period, when the oxygen source of carbon burial exceeds the oxygen sink—thus reducing materials from tectonic processes—oxygen emerges as the chemically dominant gas. By including consumers as a third ecosystem, the Archean model extends through the transition and into the Proterozoic model without loss of self-regulation. These models are stable, indifferent to the choice of initial conditions, and can resist perturbation. The values of climatic and chemical properties they predict are always realistic, a direct result of considering organisms as responsive entities that bear a limited range of tolerances to climatic and chemical variables.

Among the insights that come from a Gaian approach: Planetary life can never be sparse. A planet with sparse life could never self-regulate. The geophysical and geochemical evolution of terrestrial planets moves progressively toward states like those of Mars and Venus now. During this evolution, there will be a period when conditions are favorable for life. In this window of opportunity for newly emergent life, organisms must survive and reach a sufficient abundance to affect and couple-in with geochemical evolution; mere adaptation is not enough. If organisms fail to achieve this link, planetary conditions will continue to change inorganically until the point is reached where life is impossible.

11. See n. 8.

Another insight from Gaia is that the biota are continuously engaged in the removal of carbon dioxide from the air and that the present cool climate results from biological pumping.[12] The only long-term sink through which the carbon dioxide leaves the atomosphere is the weathering away of calcium silicate rock. Carbon dioxide and rainwater react with the rock to produce calcium bicarbonate and silecic acid, both of which are water soluble and travel along the streams and rivers to the ocean.

It used to be thought that this was just a chemical reaction, but in the real world, organisms from bacteria to trees grow on the rocks and accelerate the weathering. The extent to which organisms accelerate the rate of rock weathering by carbon dioxide was investigated by Swartzman and Volk. They found the presence of organisms increased by one thousand-fold the rate of the weathering of basalt rock by CO_2.[13] This geophysiological interpretation of the long-term regulation of climate is supported by paleobotanical observations of plants before and after the cretaceous tertiary extinction (KT).[14] They concluded that the KT event was followed by a 10°C rise in global mean temperature, a rise that persisted for between one-half and one million years. The KT event was biocidal on a global scale and could have partially disabled the CO_2 pumping. The time constant for CO_2 regulation by weathering is also about one million years. Viewed in this way, perhaps the warmth of the present interglacial implies a less abundant or less efficient biota.

Gaia theory also led directly to the identification of another possible climate-control system. In 1972, global-scale emission of dimethyl sulphide from the oceans was discovered during the search for a chemical agent of biological origin that would complete the natural sulphur cycle.[15] It was later proposed that emission of sulphur gases might be connected with climate regulation because of stratospheric sulphate aerosol's well-known ability to cause global cooling.[16] Further, dimethyl sulphide emissions from the oceans might affect climate much more powerfully through their oxidation with cloud condensation nuclei to form sulphuric and methane sulphonic acids.[17] If confirmed, this process is potentially as great a climate regulator as the carbon dioxide and methane greenhouse.

From a Northern Hemispheric human view we see glaciations as a

12. J. E. Lovelock and M. Whitfield, *Nature* 296 (1982), pp. 561–563.

13. D. W. Schwartzman and T. Volk, *Nature* 340 (1989), pp. 457–459.

14. J. A. Wolfe, *Nature* 343 (1990), pp. 153–156.

15. J. E. Lovelock, R. Maggs, and R. A. Rasmussen, *Nature* 237 (1972), pp. 452–453.

16. G. Shaw, *Climatic Change* 5 (1983), pp. 297–303.

17. R. J. Charlson, J. E. Lovelock, M. O. Andreae, and S. J. Warren, "Cloud Albedo and Climate," *Nature* 326, no. 6114 (1987), pp. 655–661.

disaster, but from a planetary viewpoint they may be a preferred state, one where life is more vigorous. The global cool comfort of an ice age ecosystem is consistent with a low level of carbon dioxide, 180 parts per million, but it requires a more powerful biological pump. The new land exposed in the tropics when sea level was over 100 meters lower than it is today must have been colonized. There also seems to have been a greater vigor in the ocean ecosystems, indicated by copious emissions of sulphur gases from them. The swings of temperature between glacials and interglacials could be interpreted, then, as the oscillations of a control system near the limits of its capacity to regulate—triggered, not caused, by the insolation changes of the Milankovitch effect. And so the current interglacial appears as a planetary fever, a pathology. Pollution of the atmosphere with greenhouse gases and the widespread destruction of natural habitats can only become further insults to an already weakened system. Forced too far, a sudden transition to a new homeostasis could occur, possibly to a much higher global temperature.

W. S. Broecker has warned that geophysical processes alone could lead to "surprises in the greenhouse."[18] As an active control system, Gaia can speed and intensify such surprises. At first, a systems response might mask the expected greenhouse warming with, for example, an actively increased cloud cover. We might then become complacent, or argue that the greenhouse effect has been overstated, only to be surprised by catastrophic climate change when the system is overwhelmed.

TESTABILITY

Gaia theory is testable, and the results are listed in the table (page 144). Why then has Gaia been so unpopular among scientists? An obvious answer is the natural inertia of science; large-scale theories are digested slowly. It took forty years for another Earth theory, plate tectonics, to be widely accepted. With Gaia there are, I think, additional reasons for its unpopularity.

Gaia is developing normally in the earth sciences and in time will either be accepted or rejected on the evidence. Some biologists, however, denounced Gaia even before the evidence began to be available: Gaia was not just wrong, it was dangerous! Postgate referred to Gaia as "silly and

18. W. S. Broecker, *Nature* 328 (1987), pp. 123–126.

Tests of Predictions Based on Gaia Theory

Prediction from Gaia	Test and Result
Mars lifeless (from atmospheric evidence, 1968)	Viking Mission (1977); strong confirmation; *footnote 2*
Organisms make compounds that can transfer essential elements from ocean to land surfaces (1971)	Dimethyl sulphide and methyl iodide found (1973); *footnote 15*
Climate regulated by the control of CO_2 through biologically enhanced rock weathering (1981)	Microorganisms greatly increase rock weathering (1989); *footnotes 12 and 13*
Climate regulated through cloud-density control linked to algal sulphur gas emissions (1987)	Still under test; *footnote 17*
Oxygen level at 21 $\pm$5 percent for past 200 million years (1973)	New hypotheses for regulation by fires; phosphorus cycling; *footnote 8 (Ages)*
Archean atmospheric chemistry dominated by methane (1988)	Still under test; *footnotes 7 (Lovelock) and 8 (Ages)*

dangerous . . . pseudoscientific mythmaking."[19] He went on to blame Gaia for the disrepute now attending science.

Even my friends among biologists become anxious when I talk of Gaia as an organism. Yet in so doing I merely continue a tradition perhaps begun with Leonardo da Vinci, and made scientific by James Hutton. Biologists rejected this tradition in the nineteenth century when their field first became intensely reductionist. Biology seems still to be imprisoned in a narrow, almost puritan, reductionism. Is it the consequence of a long and continuing war with vitalists, animists, and religious fundamentalists?, I wonder. In war, it is said, the first casualty is truth. I think that science, when it engages in adversarial encounters with crude dogma or non-science, loses objectivity and grows dogmatic itself. In their criticisms, biologists seem to see Gaia as another threat to the sanctity of neo-darwinism. No one denies the power of reduction with its apotheosis in the triumphs of molecu-

---

19. J. Postgate, *New Scientist* (April 7, 1988), p. 60.

lar biology. Reductionists' dogmatic crusade has made holism a pejorative term. Now, cybernetics, Gaia, and systems analysis are categorized with "new age" philosophy and organic food. No wonder the greens prefer Gaia to biology.

A great loss results from this rejection of a rigorous top-down approach to systems. Holism is complementary—not adversarial—to reduction. What engineer today would dare to dissect a working computer into its hardware components before testing it as a whole system? Without holistic thinking, physiology, engineering, and Gaia would be disabled.

The most difficult and the dirtiest criticism of Gaia is that there is nothing new about the theory, that it has all been said before—difficult, because there is much truth in the criticism; dirty, because, as William James said of the fate of any new idea, "First, it's absurd, then maybe, and last, we have known it all along." Most biologists and many geochemists are ignorant of the details of control theory, the domain of engineers and physiologists. Merely for them to say Gaia was explained by the balance of nature long ago is not enough. If it were, then explanations for why the concentration of oxygen in the atmosphere is 21 percent, or why the Earth's climate has remained favorable for 3.6 billion years, would already exist. Pre-Gaia, such questions were rarely asked, and would have been as pointless as asking an anatomist or a biochemist how human temperature is regulated. Such questions about systems cannot be answered by the separated disciplines of biochemistry or biogeochemistry, nor by neo-darwinist biology. The answer comes from physiology or control theory.

What of the criticism that Gaia gives support to anti-science? True, Gaia as a name, or a sign, has extended far beyond my intentions. As the semiotician Myrdene Anderson put it several years ago, "Gaia is an empty sign with near infinite capacity for signification." I watch it filling fast, and mostly with rubbish, like an empty skip left on a London street. Surely though, this is the fate of any new sign and has nothing to do with the quality of the science of Gaia. I have offered "geophysiology" as an alternative name, but so far there have been few takers.

I am not dismayed by the wide general interest in Gaia; rather, I am moved by the interest shown by theologians and philosophers. Whether or not Gaia is an accurate description of the Earth, it forces a different view from that of conventional wisdom—a view that could be crucial to understanding the consequence of our pollutions and environmental disturbances.

Gaia is ecumenical across and beyond the sciences. In recent years, organized religion has grown aware that the environmental stance appropriate in biblical times is no longer so in a world of five billion people. The Earth is more than a gift from God given simply for the benefit of humans. Now that we, and our dependent crops and livestock, occupy so many available habitats of the Earth, theologians wonder if it is right for us, as stewards, to render the whole planet a sty—no matter how hygienic and well run. Gaia offers a welcome alternative to biblical humanism. In Gaia, we are part and partners of a democratic entity. The rules are insistent: Species that harm the environment are voted out through natural selection. If we are truly concerned for mankind, we must respect other organisms. If we think of nothing but selfish human greed and ignore the natural life of the Earth, we have set the scene for our own destruction and that of the comfortable Earth we know. Just now, we seem like the Gaderene swine driving our polluting cars down to a sea rising to meet us.

Arthur Clarke's notable comment, "How inappropriate to call this planet Earth when clearly it is Ocean," illustrates the wisdom of a top-down view. Few have been privileged, as astronauts, to see the Earth in its splendour from above. But anyone can rise above dogma—scientific or religious—and look down to see and cherish a most seemly Earth.

PETER H. GLEICK

Climate Changes, International Rivers, and International Security: The Nile and the Colorado

Peter H. Gleick is the director of the Global Environment Program of the Pacific Institute for Studies in Development, Environment, and Security, in Berkeley, California. He is a member of the Climate and Water Panel of the American Association for the Advancement of Science and has a MacArthur Foundation research and writing fellowship to study the implications of climatic changes for international water resources and international security. From 1980 to 1982, he served as deputy assistant for energy and the environment to the governor of California. Dr. Gleick is an active participant in several joint U.S.–Soviet climate projects, including the innovative Greenhouse Glasnost Teleconference and the official climate program Working Group VIII. He has published extensively on the subjects of the greenhouse effect, climate change and water resources, and the links between global environmental problems and international security.

G lobal environmental problems are increasingly perceived as threats to international security and politics. Among these problems, global climatic change—the so-called greenhouse effect—appears to have the greatest potential for altering natural ecosystems, complicating international agricultural trade, reducing the availability of fresh water, and increasing the risk of border hostilities over environmental refugees. The effects of climatic changes on shared freshwater resources may be particularly severe.

Expected increases in temperature, changes in rainfall patterns, and altered frequency and severity of storms will affect the demand for and the supply and quality of freshwater resources. Such changes will be felt most strongly in regions where tensions already exist over shared rivers.

WATER AND POLITICS

Water flows without regard for political boundaries; indeed, most freshwater resources are shared by two or more nations. The immediate result of sharing water resources may be friction between users caused by differences in the way each uses its share of the water. The ultimate result may be conflict and war.

Table 1.
Number of Countries (by Continent) with
More than 75 Percent of Total Area Falling
Within International River Basins

Africa	23
Asia	8
Europe	13
North and Central America	0
South America	6

Source: United Nations, 1978.

Table 2.
International River Basins

	Number of International River Basins	Percent of Area in International Basins (%)
Africa	57	60
Asia	40	65
Europe	48	50
North and Central America	34	40
South America	36	60
TOTALS	215	47

Source: United Nations, 1978.

Table 3.
Rivers with Five or More Nations
Forming Part of the Basin

River	Number of Nations	Area (km^2)
Danube	12	817,000
Niger	10	2,200,000
Nile	9	3,030,700
Zaire	9	3,720,000
Rhine	8	168,757
Zambezi	8	1,419,960
Amazon	7	5,870,000
Lake Chad	6	1,910,000
Mekong	6	786,000
Volta	6	379,000
Elbe	5	144,500
Ganges-Brahmaputra	5	1,600,400
La Plata	5	3,200,000

Source: United Nations, 1978.

Forty-seven percent of all land area falls within international river basins, and more than 200 river basins are multinational, including 57 in Africa and 48 in Europe (see Tables 1 and 2). The extent of this interdependence can be seen in Table 3, which lists thirteen major rivers with five or more nations forming part of the watershed.

Variations in climatic conditions that alter the availability and quality of freshwater resources can affect intranational and international relationships, and disagreements over shared water resources have occurred in a number of regions around the world (Biswas, 1983; Cooley, 1984; Falkenmark, 1986; Gleick, 1988, 1989b; Hundley, 1966; Widstrand, 1980). Unless water is treated in international discussions as a resource subject to changing natural conditions, conflicts may increase in frequency and severity as climatic changes begin to appear.

Among the rivers that historically have been a source of international tension or competition over water resources are the Jordan and Euphrates in the Middle East; the Nile, Zambezi, and Niger in Africa; the Ganges in Asia; and the Colorado and Rio Grande in North America. As water demands increase with population growth and the greater demands of industry, the probability of conflict over remaining water resources will also increase (Gleick, 1989a). In some cases, international treaties have been worked out to settle disputes. In other cases, water conflicts have not yet been resolved.

Future climatic changes can reduce or exacerbate these water-related tensions. Among the critical concerns are changes in (1) water availability from altered precipitation patterns or higher evaporative losses due to higher temperatures; (2) the seasonality of precipitation and runoff; (3) flooding or drought frequencies; and (4) the demand for and supply of irrigation water for agriculture. Nowhere, however, have the complications posed by climatic changes been incorporated into international negotiations or hydrologic agreements.

Resolving the problems raised by climatic change will require a new perception about how water and climatic variability should be treated and a broadening of the general principles governing international watercourses to include the problems posed by long-term climatic changes. We can explore ways of incorporating guidelines to deal with problems caused by possible climate change into international water agreements, with a focus on two rivers: the Colorado River, shared by the United States and Mexico, and the Nile, shared by nine nations in Africa. Table 4 lists important characteristics of these two rivers.

Table 4.
Characteristics of the Nile
and Colorado Rivers

Characteristic	Colorado	Nile
Length	2,250 km	6,500 km
Nature	Mountainous origin; desert	Mountainous origin; desert
Number of nations sharing	2	9
Annual flow	18 km^3; high variability	84 km^3; high variability
Allocation	Total by 2040	Total by 2015
Principal uses	Agricultural (irrigation), urban, industrial	Agricultural (irrigation), urban, industrial
International treaty	Since 1944	Since 1959 (only 2 parties)
Water shares	Fixed allocation	Fixed allocation
Shortage procedures	Vague; difficult to implement	Vague; difficult to implement
Conflict resolution	None (permanent joint commission)	None (permanent joint commission)
Hydrologic modeling	Extensive	Extensive
Climate impact studies	Several (others ongoing)	Few (others ongoing)
Expected climate changes	Temperature: hotter Precipitation: uncertain Runoff: likely decreases	Temperature: hotter Precipitation: uncertain Runoff: uncertain
Potential for water conservation	High	High
Potential for flow augmentation	Low	Low to moderate
Potential for conflict	Medium	High

THE COLORADO AND
THE NILE

"The next war in our region will be over the
waters of the Nile, not politics."
—Egypt's Ministry of Foreign Affairs, 1985

"[Colorado River water] represents a national
interest superior to any other."
—Mexico's Secretaria de Relaciones Exteriores, 1945.

The Nile River Basin

The Nile River is an international river of tremendous regional importance.
It flows through some of the most arid parts of northern Africa and is vital
for agricultural production in Egypt and the Sudan. As a result, the river
is extensively used—so extensively almost no water reaches the Mediterranean Sea. Rational utilization of the Nile is complicated by the fact that the
catchment and riparian drainage area of the river is shared by nine nations:
Egypt, the Sudan, Ethiopia, Kenya, Tanzania, Zaire, Uganda, Rwanda, and
Burundi. Although the principal water users are Egypt and the Sudan,
almost all the runoff in the river is generated by precipitation in other
countries, particularly the highlands of Ethiopia.

The Nile's freshwater resources are widely shared and heavily used.
Among the many benefits of the Nile are: irrigation, navigation, drinking
water, and waste disposal; natural habitats for fish, wildlife, and birds; and
hydroelectricity generation. Growing population pressures and increasing
development along the river already are straining the limited freshwater
resources of the basin; future climatic changes will add other pressures with
widespread political ramifications.

Because the Nile is so important for the countries of northeastern Africa,
any changes in river flow have enormous impacts. The recent decade of
drought stands as a painful reminder of Egypt's vulnerability. Yet the
problems caused by the natural variations in Nile runoff may soon be
compounded by global climatic changes that affect evaporation, precipitation, and ultimately, water availability. These climatic changes—the
so-called greenhouse effect—must lead to a reevaluation of water allocation
and methods for resolving disputes over shared resources. Without such a
reevaluation, the risk of friction and conflict in the region will increase.

Competition for the waters of the Nile arose in the early 1900s, as Egyptian needs were growing, and the question of the allocation of the river flow was submitted to international mediation (including Indian, British, and American representatives) in 1920. In 1929, a first Nile Waters Agreement was adopted. A ratio of flow of 1:12 (Sudan: Egypt) was adopted, with total allocated flows of 52,000 million cubic meters per year.

Conflict over allocations resurfaced in the 1950s, a period of internal unrest in both Egypt and the Sudan. In 1958, Egypt dispatched an unsuccessful military expedition to reclaim disputed border territory, and in 1959, the Sudanese abrogated the 1929 agreement by unilaterally raising the height of the Sennar Dam in their country. By late 1959, however, a change in the political climate had led to reopened negotiations, and a new Agreement for the Full Utilization of the Nile Waters was signed. This raised the Sudanese allocation to 18,500 million cubic meters per year and set Egypt's quota at 55,500 million—a 1:3 ratio. In addition, a permanent joint technical commission was empowered to adjudicate future conflicts.

The agreement reached in 1959 between Egypt and the Sudan is often considered an historic example of the type of cooperation possible between riparian neighbors, yet it ignores the seven other nations. In the future, these other basin nations are likely to play a larger role in the use of the Nile because of their growing populations and irrigation demands (see Table 5). Ethiopia, for example, has already announced it reserves the right to develop hydrologic projects on the Blue Nile and in headwater tributaries to the Nile located in the Ethiopian highlands, and it has called for new basin-wide agreements to allocate shares among the other riparian nations.

Unfortunately, the 1959 treaty has a number of other shortcomings that will complicate resolving climate-related problems (Goldenman, 1989). First, water allocations are fixed, rather than proportional. Second, the permanent joint technical committee set up by the treaty has little authority, and the treaty contains no dispute-resolution procedures and no process for amendment. Third, the provision for drought-induced shortages is vague and difficult to implement. These ambiguities and omissions could result in a revival of friction if runoff in the Nile were to be reduced by climatic changes, or if the frequency and intensity of flooding in the basin were to increase. In fact, there is some evidence that the greenhouse effect could lead to *both* a reduction in overall stream flow in this region *and* an increase in the intensity of seasonal floods (Gleick, 1989c).

Recent climatic extremes in the Nile basin provide insights into the vulnerability of water users in the basin. Twice during the last twenty years, precipitation levels dropped in the headwater regions, and severe droughts

Table 5.
Demographic Details: The Nine Nile Basin Countries

Country	Area (Thousands km^2)	1989 Population (Millions)	Population Growth Rate (% Yr)	Irrigated Area Arable Cropland (10^6 Hectares)	Total Arable Cropland (10^6 Hectares)
Egypt	1,000	51	2.3	2.5	2.5
Sudan	2,376	24	2.9	1.9	12.5
Ethiopia	1,101	49	2.8	0.14	13.9
Uganda	200	18	3.5	0.01	6.6
Kenya	569	24	4.2	0.05	2.4
Tanzania	886	26	3.7	0.1	5.2
Burundi	26	5	2.8	0.07	1.3
Rwanda	25	7	3.4	<0.01	1.0
Zaire	2,268	34	3.0	<0.01	6.6

1 hectare = 2.38 feddans = 2.47 acres
1 feddan = 1.038 acres = 0.42 hectares
100 hectares = 1 km^2

Source: World Resources Institute, 1988.

resulted. In 1987–1988, water levels in the Nile were at their lowest in recorded history and Egypt was unable to draw its entire allotted share (Ross, 1987). It is not yet possible to say with certainty whether these anomalies are examples of short-term weather variability or the first indications of longer-term persistent changes. What is known, however, is that these droughts led to significant hardship in lower basin states. They also showed the inadequacy of plans to allocate such shortages: the technical committee, which was given authority in the 1959 treaty to draw up plans for allocating drought shortages, failed to do so.

The fate of the Nile is of immense importance to both Egypt and the Sudan, where it presently sustains a population of 75 million people. Yet all of the nine nations that share the river could influence its flow and quality. Each of them, but particularly Egypt and the Sudan, assumes that the Nile is the one reliable element in the increasingly complicated challenges that face them. Any climatically induced change in water availability threatens this assumption and will contribute to political jousting. The nine basin states need to resolve ambiguities over the allocation of shortages and reach a specific agreement about how climatic changes are to be identified and handled before such changes appear.

The Colorado River Basin

The southwestern United States shares more than a common border and history with Mexico—it also shares limited water resources. Because of the arid nature of this region, these water resources have been the focus of long and sometimes bitter controversies among farmers, among states, and between the United States and Mexico. Although past disagreements have been mostly resolved, future climatic changes that adversely affect the existing hydrologic regime of the basin could contribute to U.S./Mexico tensions. Any persistent shortage to Mexico is certain to cause international problems, even if the shortage is shared according to existing provisions. Given the recent difficulties between the two nations over drug traffic, immigration laws, Central American politics, and issues like the transborder pollution from copper smelters, a resurgence of problems over water resources would be cause for concern.

The Colorado River flows over 2,000 kilometers from its headwaters in the Rocky Mountains to its delta on the Gulf of California in Mexico. The Colorado is the principal river system in the southwestern United States and northwestern Mexico, moving through regions that are extremely dry and hot. Population growth and the rising demand for irrigation water have led to a nearly total allocation of Colorado River waters. In many years no flow reaches the Gulf of California at all, just as the waters of the Nile now rarely reach the Mediterranean.

The heavily exploited water resources of the Colorado River basin play a crucial role in the economic and agricultural development of important regions of both the United States and Mexico. Two major areas of irrigated agriculture in Mexico use Colorado River water—the Mexicali Valley and the San Luis Valley near the delta. In 1988, the Colorado met the domestic needs of nearly a million Mexicans (Holburt, 1978), and demands are growing rapidly. Water from the Colorado is also used to irrigate the Imperial Valley in California and to supply drinking water for many urban centers in the Southwest.

Controversy between the United States and Mexico over water began in the 1870s, when irrigation development that reduced water supplies and quality brought open conflict (Hundley, 1966). Tensions grew as low-water years and more upstream diversions decreased flows. Finally, in 1896, after a series of diplomatic complaints, legal rulings, and political maneuvers, the United States directed the International Boundary Commission to study the problem of equitable distribution of the Colorado.

In February 1944, after nearly half a century of debate and controversy, the United States and Mexico signed a treaty guaranteeing 1.5 million acre-feet (1.8 billion cubic meters) of Colorado River water to Mexico annually. The long and careful negotiations leading to the 1944 treaty failed to resolve two important problems: (1) guaranteeing the quality of the water to be delivered to Mexico; and (2) the possibility of long-term reductions in water availability. The first problem strained U.S.–Mexican relations beginning in the 1960s, when U.S. irrigation projects caused significant increases in salinity levels near intakes for the Mexican supplies. In 1961, Mexicans were forced to let their allocation flow unused to the Gulf of California, rather than risk damage to their soils by using the highly saline water. After protests by the Mexican government, water-quality discussions were opened between the two countries. These led eventually to Minute 242 of the International Boundary and Water Commission, which set a limit on the salinity of water delivered to Mexico. The agreement was followed several years later by legislation in the United States to implement measures for the control of salinity in the Colorado.

The second problem—a shortfall in water quantity—was addressed in the 1944 treaty but in a very limited way. In the event of an "extraordinary drought" or a "serious accident" to the irrigation system in the United States, the water allocated to Mexico is to be reduced in the same proportion as consumptive uses of water in the United States are reduced. Neither of these terms was defined by the treaty, and the negotiating record offers little clarification. Because Mexico has consistently received all the water allocated to it by the 1944 treaty, this provision has yet to be tested.

Three situations could lead to periodic shortages in the United States and, perhaps, Mexico: (1) the growth of consumptive water demand to or beyond the limits of the present natural supply; (2) a persistent but temporary drought; and (3) a change in the climatic conditions that permanently reduces the overall availability of water.

A specific example can demonstrate the severity of the problem posed by an increased demand for water in the Colorado River basin. The U.S. Bureau of Reclamation estimates by the year 2000, consumption of water and reservoir evaporation in the upper and lower basins will total nearly 14.5 million acre-feet (18 billion cubic meters) annually, with an additional

1.5 million acre-feet (1.8 billion cubic meters) per year delivered to Mexico, for a total annual commitment of nearly 16 million acre-feet (nearly 20 billion cubic meters) (U.S. Bureau of Reclamation, 1982, 1984). If there is initially no drawdown of the reservoirs, runoff to Mexico will equal runoff into the Colorado River less consumptive uses and evaporative losses in the United States.

If the natural flows originating in the upper basin average 15 million acre-feet (18.5 billion cubic meters) per year and if consumptive usage reaches the levels described above, the annual delivery of water to Mexico as specified by the treaty would be threatened periodically, as early as the mid–1990s. If the presently unquantified water rights of the Navajo Indians are resolved, problems could arise even sooner (Kneese and Bonem, 1986). Indian water rights could entail several million acre-feet per year.

The large storage capacity in the Colorado River basin is adequate to delay a crisis for several years, but it provides no long-term solution to overallocation or permanently reduced flows. During the early years of shortage, the large Colorado reservoirs could, theoretically, contribute enough water to make up the difference and meet the Mexican allocation. After some reservoir depletion, however, additional water demand would begin to threaten the generation of hydroelectric power, long-term water supply, and recreational uses. Moreover, if the long-term average runoff in the upper basin is actually closer to Stockton and Jacoby's (1976) estimate of 13.5 million acre-feet per year—10 percent less than the 15 million estimated by the Bureau of Reclamation—the assistance provided by even the largest reservoirs ultimately disappears.

The agreements between the United States and Mexico were completed with the implicit assumption that the average availability of water in the Colorado River basin would continue to be the same in the future as it was in the recent past, that is, that climate would remain stationary. If this assumption is wrong, the implications for water availability when considering future changes in global and regional climatic conditions are likely to be even more important than the problems raised by existing climatic variability.

Anthropogenic climatic changes (changes to climate caused by human activities) will alter the availability of water in the Colorado River through adjustments in either the timing or the magnitude of runoff and water demand. Although water availability in some regions may increase if precipitation increases, there are indications that the Colorado River and other low- and middle-latitude arid and semiarid regions will not be so lucky (Flaschka, Stockton, and Boggess, 1987; Gleick, 1986, 1987; Manabe and

Wetherald, 1986; Manabe, Wetherald, and Stouffer, 1981; Revelle and Waggoner, 1983; Stockton and Boggess, 1979; Nash and Gleick, 1990). Table 6 lists estimates of plausible changes in annual runoff in the Colorado River basin and similar semiarid basins that would be caused by future climatic changes. Some of these studies predict annual runoff will decrease even if precipitation increases (Stockton and Boggess, 1979; Revelle and Waggoner, 1983), and there is no question that temperature increases alone could greatly reduce the availability of water in semiarid basins (Flaschka et al., 1987; Gleick, 1986, 1987; Nash and Gleick, 1990).

As noted earlier, during severe droughts Mexico's Colorado River allocation can be decreased proportionately "as consumptive uses in the United States are reduced." Unfortunately, the interpretation of these provisions of the 1944 treaty remains open to debate because treaty negotiators made no attempt to agree on the meaning of "extraordinary drought" or on a system to decide when such a drought was in progress. Ambiguity over these fundamental matters will almost certainly lead to problems in coming years.

RESPONDING TO CLIMATE CHANGE

Important international policy questions arise when we consider the possibility of climate-induced shortages in shared river basins. What, for example, are the implications for the treaty obligations of the United States toward Mexico, or the Sudan toward Egypt, of a reduction in runoff due to climatic changes? Careful thought must be given also to the type of policy responses that should be taken to prevent the worst and most likely consequences of such changes. A reduction in annual runoff in the Colorado or Nile of only 10 to 20 percent—well within the range of estimates of recent studies—would decrease available water by a large absolute amount. Since the small margin of unused water in these regions would not cover persistent shortfalls of this magnitude, prudent planning would have to explore the implications of such a situation. Table 6 suggests that even more extreme decreases in runoff are possible: a permanent reduction in riverflow of 30 to 50 percent would cripple industries, agriculture, and communities throughout both regions. Domestic and international competition for remaining water would certainly ensue.

A distinction must be made between short-term and long-term shortfalls. Short-term droughts rarely persist longer than a year or two. Long-term water shortages caused by natural variability or anthropogenic climate

Table 6.
Effect on Runoff of Climatic Changes in
Major Semiarid River Basins

Author/Basin	Change in Annual Runoff (%)
Stockton and Boggess (1979)	
Upper Colorado River basin	
P −10%; T +2°C	−35
P +10%; T +2°C	−33
Lower Colorado River basin	
P −10%; T +2°C	−56
P +10%; T +2°C	− 2
Nemec and Schaake (1982)	
Pease River basin, Texas	
P −10%; T +1°C (ET +4%)	−50
P +10%; T +1°C (ET +4%)	+50
P −10%; T +3°C (ET +12%)	−50
P +10%; T +3°C (ET +12%)	+35
Revelle and Waggoner (1983)	
Colorado River basin	
P −10%; T +2°C	−40 ± 7.4
P +10%; T +2°C	−18
Flaschka et al. (1987)	
Great Basin subbasins	
(primarily Utah to Nevada)	
P −10%; T +°2C	−17 to −28
P +10%; T +°2C	+20 to +35
Gleick (1986, 1987)	
Sacramento River basin	
P −10%; T +2°C	−18
P −20%; T +2°C	−31
P +10%; T +2°C	+12
P +20%; T +2°C	+27
P −10%; T +4°C	−21
P −20%; T +4°C	−34
P +10%; T +4°C	+ 7
P +20%; T +4°C	+23

P = precipitation; T = temperature; ET = evapotranspiration.*
*Evapotranspiration equals evaporation plus transpiration (which is the total use of water
by plants).

changes would, effectively, have time scales on the order of decades to centuries. Because the climate is assumed to be constant, water-resource systems and institutions are designed to accommodate only short-term fluctuations in water availability. Yet even short-term droughts can have drastic local impacts. The 1976–1977 drought in the western United States resulted in extensive forest damage, shifts in agricultural productivity, livestock losses, and changes in the operation of reservoirs and aqueducts. If this unusual drought had persisted for one more year, the impacts would have been devastating. The drought in northeastern Africa during the 1980s had similar effects on hydroelectricity production, agricultural production, and urban development.

Despite recent advances in the codification of agreements to govern international rivers, there is still no consolidated international water law. Nor are there clear precedents that take into account the problems that will be raised by changing climate. Thus, future alterations in the availability of water resulting from changes in climate are likely to exacerbate existing tensions over water resources, manifested in both regional and international disputes.

While international water law does not explicitly address the problem of changing climatic conditions, some international legal principles are at least beginning to take climate into account as an important factor. For example, the Helsinki Rules on the Uses of the Waters of International Rivers, formulated in 1966 and revised in 1986, specify that "each basin State is entitled, within its territory, to a reasonable and equitable share in the beneficial uses of the waters of an international drainage basin." The Rules set forth conditions for determining what is "reasonable and equitable," one of which is the "climate['s] affecting the basin" (International Law Association, 1967, revised 1986). This provision alone, however, will not resolve problems posed by climatic changes: it lacks specificity, and the Helsinki Rules are not universally accepted as a binding guide to international water problems.

The International Law Commission is considering new language and principles for shared international river systems without specific treaties or agreements. Global warming and climatic change are explicitly discussed in the Commission's 1989 report (McCaffrey, 1989). Although no specific measures are suggested, climatic change is considered likely to increase the risks of hydrologic changes and conflict, and is therefore more reason for nations to cooperate in preparations for water-related disputes.

In the context of international treaties, the possibility of water shortages under conditions of persistent climatic changes should be the subject of discussions among concerned parties. The question of what constitutes a persistent change should be answered. Two new treaty components may be necessary: (1) a clear definition of "climatic change," and a methodology for distinguishing between long-term and short-term events; and (2) unambiguous allocation of shortages. The flexibility of treaties must be improved to account for both short-term droughts and long-term climatic changes. To wait until serious pressures on the water resources of the Colorado and Nile develop will only increase the difficulty of resolving them.

CONCLUSIONS

The challenge of global climatic change makes multinational coordination in the use of shared freshwater resources a necessity. If we wish to avoid future conflicts over limited water resources, national self-interest and self-preservation require that instead of selfishly controlling and restricting access to common resources, we must develop mechanisms for equitable sharing on a global scale.

Early conflicts over the waters of the Nile and the Colorado were resolved mostly by international treaty, but none of the existing treaty provisions addresses the possibility of problems caused by climatic changes. With the pressures on the limited resources of both rivers and the complication of climate-induced shortages, an effort should be made to include allowances in the two treaties for the effects of future climatic changes. Frictions between users could be resolved within the framework of existing agreements through treaty renegotiation or clarification. Unfortunately, the very act of reopening discussion is fraught with intense political, economic, and social difficulties, both international and domestic.

In the short run, the most appropriate political and economic action is to reduce inefficient consumption of water. Such reduction will provide a cushion for meeting treaty obligations and add flexibility to operating strategies for existing water systems. In the long run, however, the problem of growing demand for shared waters and the possibility of climate-induced reductions in flows of international river basins must be addressed directly. General principles governing international watercourses must be broadened and climatic changes explicitly incorporated into long-range water-resource planning. Because of the importance of shared water resources in regions throughout the world, and because of the severity of possible future climatic changes, conflict over water resources will continue to develop unless the problems raised by climatic change can be identified and resolved.

The alternative: Do nothing and assume the risks of current management of international river resources are smaller than the problems that might arise with an attempt to develop a better system. This argument has two serious flaws. First, even in the absence of concern over climatic changes, there are regions with current serious disputes over water resources. This alone suggests the need for more comprehensive agreements. Second, the magnitude of likely climatic changes could dwarf natural variations in the basins in question. Unless an effort is made to develop nonconfrontational solutions to water scarcity and multinational allocations, eventual conflict seems unavoidable.

REFERENCES

Biswas, A. K., "Shared Natural Resources: Future Conflicts or Peaceful Development?" in R. J. Dupuy (ed.), *Settlement of Disputes on the New Natural Resources.* The Hague: Martinus Nijhoff, 1983, pp. 197–215.

Cooley, J. K., "War over Water," *Foreign Policy* 54 (1984), pp. 3–26.

Falkenmark, M., "Fresh Waters as a Factor in Strategic Policy and Action," in A. H. Westing (ed.), *Global Resources and International Conflict: Environmental Factors in Strategic Policy and Action.* New York: Oxford, 1986, pp. 85–113.

Flaschka, I. M., C. W. Stockton, and W. R. Boggess, "Climatic Variation and Surface Water Resources in the Great Basin Region," *Water Resources Bulletin* 23, no. 1 (1987), pp. 47–57.

Gleick, P. H. "Methods for Evaluating the Regional Hydrologic Impacts of Global Climatic Changes," *Journal of Hydrology* 88 (1986), pp. 97–116.

———, "Regional Hydrologic Consequences of Increases in Atmospheric CO_2 and Other Trace Gases," *Climatic Change* 10 (1987), pp. 137–161.

———, "The Effects of Future Climatic Changes on International Water Resources: The Colorado River, the United States, and Mexico," *Policy Science* 21 (1988), pp. 23–39.

———, "Climate Change and International Politics: Problems Facing Developing Countries," *Ambio* 18, no. 6 (1989a), pp. 333–339.

———, "The Implications of Global Climatic Changes for International Security," *Climatic Change* 15, no. 1-2 (1989b), pp. 309–325.

———, "The Vulnerability of Runoff in the Nile Basin to Climatic Changes." International Seminar on Climatic Fluctuations and Water Management, December 11–14, 1989, Cairo. Berkeley: Pacific Institute for Studies in Development, Environment, and Security, 1989c (No. A-15).

Goldenman, G., *International River Agreements in the Context of Climatic Changes.* Berkeley: Pacific Institute for Studies in Development, Environment, and Security, 1989.

Holburt, M. B., "International Problems," in D. F. Peterson and A. B. Crawford (eds.), *Values and Choices in the Development of the Colorado River Basin.* Tucson: University of Arizona, 1978, pp. 220–237.

Hundley, N., Jr., *Dividing the Waters: A Century of Controversy Between the United States and Mexico.* Berkeley and Los Angeles: University of California, 1966.

International Law Association, "The Helsinki Rules on the Uses of the Waters of International Rivers." 52nd Conference of the International Law Association, Helsinki, 1967 (revised 1986).

Kneese, A. V., and G. Bonem, "Hypothetical Shocks to Water Allocation Institutions in the Colorado Basin," in G. D. Wetherford and F. L. Brown (eds.), *New Courses for the Colorado River: Major Issues for the Next Century.* Albuquerque: University of New Mexico, 1986, pp. 87–108.

Manabe, S., and R. T. Wetherald, "Reduction in Summer Soil Wetness Induced by an Increase in Atmospheric Carbon Dioxide," *Science* 232 (1986), pp. 626–628.

Manabe, S., R. T. Wetherald, and R. J. Stouffer, "Summer Dryness Due to an Increase of Atmospheric CO_2 Concentration," *Climatic Change* 3 (1981), pp. 347–386.

McCaffrey, S., *Fifth Report on the Law of the Non-Navigational Uses of International Watercourses.* International Law Commission (41st Session), 1989. (United Nations Document No. A/CN 4/421.)

Nash, L. L., and P. H. Gleick, "The Sensitivity of Streamflow in the Colorado River Basin to Climatic Changes," *Journal of Hydrology* (on press).

Nemec, J., and J. Schaake, "Sensitivity of Water Resource Systems to Climate Variation," *Hydrologic Sciences Journal* 27, no. 3 (1982), pp. 327–343.

Revelle, R. R., and P. E. Waggoner, "Effects of a Carbon Dioxide–Induced Climatic Change on Water Supplies in the Western United States," in National Academy of Sciences, *Changing Climate.* Washington, DC: National Academy, 1983, pp. 419–432.

Rind, D., "The Doubled CO_2 Climate and Future Water Availability in the United States." (Paper submitted to *Journal of Geophysical Research,* 1987).

Ross, M., "Parched Egypt Watches Anxiously as Waters Behind Aswan Dam Recede," *Los Angeles Times,* December 26, 1987.

Secretaria de Relaciones Exteriores, "El Tratado de aguas internacionales, celebrado entre Mexico y los Estados Unidos el 3 de febrero de 1944." Mexico City: Secretaria de Relaciones Exteriores, 1947, p. 38.

Stockton, C. W., and W. R. Boggess, "Geohydrological Implications of Climate Change on Water Resource Development." Fort Belvoir, Va.: U.S. Army Coastal Engineering Research Center, 1979.

Stockton, C. W., and G. C. Jacoby, Jr., "Long-Term Surface-Water Supply and Streamflow Trends in the Upper Colorado River Basin Based on Tree-Ring Analyses." Lake Powell Research Project, Laboratory of Tree-Ring Research, University of Arizona, Tucson, 1976.

United Nations, *Registry of International Rivers.* Oxford, England: Pergamon, 1978.

U.S. Bureau of Reclamation, "Colorado River System Consumptive Uses and Losses Report 1976–1980." Denver: U.S. Bureau of Reclamation, 1982.

———, "Colorado River Simulation System (CRSS): Data Files." Denver: U.S. Bureau of Reclamation, 1984.

Widstrand, C., ed., *Water and Society. Vol. 2: Water Conflicts and Research Priorities.* Oxford, England: Pergamon, 1980.

World Resources Institute, *World Resources 1988–89.* Washington, DC: World Resources Institute, 1988.

PAUL R. EHRLICH
AND
ANNE H. EHRLICH

Population and the Greenhouse Warming

Paul R. Ehrlich is Bing Professor of Population Studies at Stanford University, where he has been a member of the faculty since 1959. He is the author of more than 500 scientific papers and articles for the popular press, and has thirty books to his credit, including: *The Population Bomb*, *The Machinery of Nature*, and *New World/New Mind*. Dr. Ehrlich has worked extensively in the area of population biology, and with his wife, Anne Ehrlich, in policy research on human ecology.

Dr. Ehrlich's fieldwork has carried him to all continents, from the Arctic and the Antarctic to the tropics, and from high mountains to ocean floor. He is a member of the National Academy of Sciences and the American Philosophical Society, and a fellow of the American Association for the Advancement of Science and of the American Academy of Arts and Sciences. He is honorary president of Zero Population Growth, Inc., a past president of the Conservation Society, president of Stanford's Center for Conservation Biology, and 1990 president of the American Institute of Biological Sciences.

Anne Ehrlich is a senior research associate in the Department of Biological Sciences at Stanford University. She is coauthor of more than a half dozen books, including *Extinction* and *The End of Affluence* (both with Paul Ehrlich), and is chair of the Sierra Club's national committee on Military Impacts on the Environment. Anne Ehrlich is an honorary fellow of the California Academy of Sciences, and was selected for the United Nations' Global 500 Roll of Honour for Environmental Achievement. She has an honorary doctorate from Bethany College.

The Ehrlichs' latest collaboration, *The Population Explosion*, has just been published by Simon & Schuster.

A recent reawakening of concern over environmental problems raises the hope that, after a decade of neglect, Americans may once again assume leadership in seeking solutions to the deepening human predicament. The size and growth of the human population play a critical role in shaping that predicament, yet the connections between its various aspects and overpopulation are seldom explicitly recognized. For instance, many population-related topics facing civilization failed to enter significantly into policy debates during the bitterly fought U.S. presidential election of 1988. These topics included:

1. The greenhouse warming, with its potential for causing severe disruption of agriculture (leading to reduced production and less dependable harvests), rising sea levels, and other serious problems.

2. High rates of soil erosion and the depletion and despoliation of groundwater supplies, both of which also have serious implications for agriculture.

3. The accelerating "extinction crisis," in which the disappearance of populations and species of other organisms threatens the provision of crucial life-support services by ecosystems and exacerbates both global warming and soil erosion.

All of these problems are caused at least partly by overpopulation in rich nations and by overpopulation and continuing rapid population growth in poor nations. Yet the population component too seldom enters discussions of global environmental problems. It was ignored in the election, it is rarely mentioned in the media, and it was difficult even to keep before the Sundance Greenhouse/Glasnost symposium.

How is population connected to greenhouse warming? The connection can be illustrated with a few simple thought experiments. It is widely recognized that industrialized countries (with about a fifth of the world's population) bear the heaviest responsibility for injecting carbon dioxide (CO_2) into the atmosphere. Roughly four-fifths of the CO_2 released by burning fossil fuels comes from automobiles, power plants, and other industrial apparatus used mainly by the rich. In terms of carbon dioxide per unit of energy generated, coal burning is the worst offender; for every ton of high-quality coal burned, roughly 3 tons of CO_2 are released. Burning natural gas produces about half as much CO_2 per unit energy as coal, and oil falls between them.

Suppose the United States decided to take dramatic steps to cut its contribution to the CO_2 component of global warming simply by terminating all burning of coal (which now supplies almost a quarter of our annual energy consumption) but not replacing it with another carbon-based fuel.

First, let us look at the consequences of large human numbers in the absence of further growth. What if China managed to keep its population at today's level, some 1.1 billion people? (It is thought by some to be around 1.2 billion already, and current demographic projections indicate it will reach at least 1.5 to 1.7 billion before growth halts.) Imagine also that China would be willing to scale back its development plans in order to help reduce the warming and only double its per capita commercial energy consumption (to about 14 percent of present per capita use in the United States, a level now found in Algeria) using its vast stocks of coal. That small step by China in moving toward what virtually everyone would agree is a legitimate development goal would add more than enough CO_2 to the atmosphere to offset the reduction of CO_2 input produced by America's unprecedented sacrifice. Thus, even without considering population *growth* in either rich or poor countries, the vast levels of overpopulation already achieved by humanity can amplify small and reasonable per capita changes in energy use into gigantic impacts.

If one does consider population growth, the situation looks even worse. If India's per capita energy consumption rose just to about the level of China's today, that combined with India's projected population growth (if one assumes some success with family-planning efforts) by the end of the next century would produce an impact equivalent to that of doubling China's per capita energy consumption without increasing its population. Even if over the next thirty-five to forty years India managed to lower its average family size from 4.3 to 2.2 children (replacement level), the country's present population of about 850 million would expand to about 2 *billion* before growth halted a century from now.

Consequently, although poor nations are now relatively minor contributors to the CO_2 load generated by burning fossil fuels, the effects of their legitimate aspirations to develop, multiplied by their population growth, will almost certainly boost their share of CO_2 releases substantially in the near future.

Other contributions of population growth to the carbon dioxide problem are also very impressive but much more difficult to quantify. Plants take up carbon dioxide in the process of photosynthesis; when they die and decay or are burned, they release it again. Today, the rate of global deforestation so overshadows that of regrowth that deforestation is a major component— estimated at 25 to 40 percent—of the total human-caused CO_2 releases to the atmosphere.

Obviously, then, one long-term measure that would help with the greenhouse problem is to regenerate forests, preferably using trees that grow slowly and whose wood when harvested would be preserved (used in construction or furniture) rather than burned—ideally, high-quality hardwoods in the tropics. But of course it is precisely in the tropics that expanding human populations are contributing heavily to forest destruction (although much of that destruction can be traced to overconsumption and overpopulation in rich nations).

In the next century, methane could surpass carbon dioxide as the most important trace gas in causing greenhouse warming. Methane is some twenty to thirty times more effective in blocking heat radiation than is CO_2, and its concentration in the atmosphere is rising much faster. The population connection here is very clear, because major sources of methane include rice paddies, the digestive tracts of cattle, landfills, and smoldering logs and exposed soils where tropical forests have been cleared and burned. All of these sources are intimately connected to the size of the burgeoning human population, so substantial reductions in methane emissions will not be very easily achieved.

Still more potent greenhouse gases are chlorofluorocarbons (CFCs), well known for their ability to destroy stratospheric ozone. One prominent atmospheric scientist has pointed out that the action taken by the United States in the 1970s to ban CFCs in aerosol sprays may have postponed the arrival of global warming by twenty years. CFCs too have a population connection; their use as refrigerants relates to population size and to the potential for both development and population growth in poor nations. Just consider the potential additions as every household in China, India, and the rest of Asia, and in Latin America and Africa acquires a refrigerator! But refrigerators are important contributors to public health and well-being; so some way must be found for people everywhere to have affordable ones without the release of CFCs into the atmosphere.

CFCs, unlike the other major greenhouse gases, present an opportunity for abatement without addressing the knottier problems of overpopulation or overconsumption. Reducing CO_2 emissions substantially and for the long term, for instance, will not be possible without tackling both the rate of fossil-fuel burning and population growth. Even with stringent energy conservation measures in place and large reductions in per capita consumption, continued population growth (as illustrated by the example of India above) would soon cancel that gain.

Similarly, success in reversing CO_2 emissions caused by tropical deforestation depends on ending population growth as well as changing land-use policies. Perhaps most of all, curbing methane releases will depend on success in population control, changes in consumptive patterns (of meat and all the materials that end up in landfills), and again, forest preservation.

The dramatic threats to civilization posed by the buildup of greenhouse gases in the atmosphere have refocused attention on increasing environmental deterioration around the world. Ironically, they also spotlight the dilemma we were already facing, as humanity has become increasingly dependent on spending "capital" to support its burgeoning numbers, rather than subsisting on its solar "income." The same burning of fossil fuels that puts carbon dioxide into the atmosphere is the most obvious case of capital-burning; others are the accelerating losses of irreplaceable soils to erosion, the mining of Pleistocene groundwater, an essentially irreversible desertification of arid and semiarid lands everywhere, and of course, destruction of natural ecosystems, especially tropical forests and the still unmeasured wealth of biotic diversity they contain.

How soon can human population growth be humanely halted? Can societies accept the necessity of a slow decline thereafter to a population size that is sustainable in the long term? Such a population must be able to support itself predominantly on income (primarily, that is, on the energy supplied daily by the sun) rather than capital (nonrenewable resources). How might this optimum population size be determined, let alone attained?

Because of the momentum for growth inherent in the youthful age composition of exploding populations, reaching a stationary population (zero population growth) on a worldwide basis will take many decades—unless some catastrophe intervenes to raise death rates substantially.

Enormous variation in growth rates prevails among nations today. A dozen or so European countries have already stopped growing or are very close to it; some have even begun slowly shrinking. By contrast, populations in most developing nations are still expanding rapidly—some by more than 3 percent per year, a rate that, if continued, would double them in twenty-three years or less. Even though some reductions in birthrates have occurred in the last two decades, most of these countries are projected by demographers to double and even triple their present populations before growth ends.

Furthermore, the assumption underlying the projections that birthrates will continue to drop in developing nations may well be faulty. Some countries that achieved substantial early success with their family-planning programs seem unable to reduce fertility below an average of 3.5 to 4 children per family, a rate that guarantees continued growth indefinitely.

Very poor people generally are unresponsive to the idea of family limitation (except for spacing of births to protect the health of mothers and infants) as long as children are seen as economically advantageous and necessary for their own security in old age. When children are economic burdens (requiring schooling and support rather than contributing to family income), old-age security is provided, and infants have a good chance of survival, parents are open to family planning. The recent lag in birthrate reduction may reflect in part a failure of some nations to establish this minimal level of family well-being for all segments of society.

Even so, the principal problem in developing nations today is not opposition to the idea of family limitation, it is a shortage of information and means: contraceptives and the ability to use them. In the past decade, the connection between overpopulation and environmental degradation, especially the deterioration of farmlands and loss of fuelwood resources, has become crystal clear to decision makers in poor countries. But at the same time, the Reagan administration sharply cut back U.S. funding for family-planning assistance to poor nations, effectively abdicating our nation's former leadership role. Although other rich nations have increased their contributions to population programs to compensate, needs have risen faster than donations. Clearly, a major factor in the failure of many poor nations to maintain and continue the birthrate declines of the 1970s is a huge and rising—and unmet—demand.

Ironically, while nearly all poor countries have more or less explicit population policies and active family-planning programs to help couples limit births, most of the relatively slow-growing industrial nations have no explicit population policies (although most do provide family-planning assistance for poor families). And the role of rich nations in generating global environmental problems makes it imperative that they limit their population sizes.

What policy options are available today to hasten the process of ending population growth? In the United States, obviously, there is a need for increased public education about population issues, for without public understanding and pressure on decisionmakers, no policies are likely to be formulated. The taboo that now seems to prevent most politicians from speaking out on population, even when they are informed about it, must be broken. In addition, the news media should be encouraged to improve its coverage of population and environmental matters.

Other sources of the all too pervasive taboo against discussing the role of population in generating civilization's problems include religious attitudes, a basic humanistic feeling that there can never be too many people, fear that paying attention to population control will divert attention from other pressing problems, and the notion that population growth is necessary to fuel economic growth.

Regardless of the sources of the taboo, we believe the chances of breaking it would be enhanced greatly if the scientific community itself were both more knowledgable about the population-resource-environment predicament and more willing to speak out about it. Among those who have begun to address the public are ecologists and evolutionary biologists. The Club of Earth, a group whose members belong to both the National Academy of Sciences and the American Academy of Arts and Sciences, released a statement to the press in September 1988, which stated in part:

> Arresting global population growth should be second in importance only to avoiding nuclear war on humanity's agenda. Overpopulation and rapid population growth are intimately connected with most aspects of the current human predicament, including rapid depletion of nonrenewable resources, deterioration of the environment (including rapid climate change), and increasing international tensions.

Civilization's problems would not all be solved if growth of the human population were halted humanely by limiting births. They would not necessarily be solved if the world population were ultimately reduced to a more or less permanently sustainable size. Society might still be plagued by racism, sexism, religious prejudice, gross economic inequity, threats of war, and serious environmental deterioration. But without population control, none of these problems can be solved; halting growth and then moving toward lower numbers simply would give humanity an opportunity to grapple with them. The old saying still holds: *Whatever your cause, it's a lost cause without population control.*

What sorts of policies should be considered? First, a commission on global population, resources, and environment might be established to generate a new public debate. The argument will be made, of course, that past studies have had little impact on policy. In 1972, the letter transmitting the report of the Commission on Population Growth and the American Future to President Richard Nixon and the Congress concluded that "no substantial benefits would result from further growth of the Nation's population, rather . . . the gradual stabilization of our population through voluntary means would contribute significantly to the nation's ability to solve its problems." President Nixon and his administration ignored the report for the most part, and the American population has grown by some 40 million since then.

Almost a decade later, the transmittal letter of *The Global 2000 Report to the President,* initiated by President Jimmy Carter, was even more explicit:

> Our conclusions . . . indicate the potential for global problems of alarming proportions by the year 2000. Environmental, resource, and population stresses are intensifying and will increasingly determine the quality of human life on our planet. . . . The earth's carrying capacity— the ability of biological systems to provide resources for human needs—is eroding. . . . Changes in public policy are needed around the world before problems worsen and options for effective action are reduced.

Global 2000 was the result of a path-breaking effort that involved coordinating the analytic capabilities of a dozen government agencies and the World Bank, with advice from people outside of government. Its "pessimistic" conclusions were undoubtedly too optimistic. One reason for this was recognized in the report itself; despite its efforts, standard problems that plague all narrow examinations of the future had not been entirely avoided by *Global 2000.* For example: "Most of the quantitative projections simply assume that resource needs in the sector they cover—needs for capital, energy, land, water, minerals—will be met. . . . It is very likely that the same resources have been allocated to more than one sector."

Unfortunately, *Global 2000* came out shortly before Ronald Reagan took office. His administration, in keeping with its antienvironmental stance and broad opposition to family planning and women's rights, did its best to sweep the report under the rug.

So the results of these earlier commissioned studies might seem to be minimal, but it is easy to underestimate their influence. A great deal of publicity accompanied each, and both provided substantial "authoritative" documentation for scientists and politicians to cite in attempting to turn knowledge into policy. *Global 2000,* moreover, had substantial impact outside the United States. It was circulated worldwide by the Carter administration through U.S. embassies, and it is credited with helping to increase awareness abroad of global environmental problems and interest in accepting family-planning aid.

With the Reagan administration's rejection of interest in the human predicament, leadership passed to the international community. In 1987, the report of the World Commission on Environment and Development, *Our Common Future,* was published. The twenty-one commissioners, all political leaders from as many nations, clearly recognized the role of population growth in shaping environmental and development problems and the need for action to curb it: "Urgent steps are needed to limit extreme rates of population growth. Choices made now will influence the level at which population stabilizes in the next century within a range of 6 billion people."

In its conclusion, the Commission noted: "Over the course of this century, the relationship between the human world and the planet that sustains it has undergone a profound change. When the century began, neither human numbers nor technology had the power to radically alter planetary systems. As the century closes, not only do vastly increased human numbers and their activities have that power, but major, unintended changes are occurring in the atmosphere, in soils, in waters, among plants and animals, and in the relationships among all of these. . . . We are unanimous in our conviction that the security, well-being, and very survival of the planet depend on such changes [those recommended by the Commission], now."

In recent years, people everywhere have become much more aware of connections between population growth and environmental problems, especially those such as desertification, tropical deforestation, and greenhouse gases. In the United States, the public seems newly receptive to the notion that demographic factors will be crucial in shaping the future of civilization.

A new study, perhaps conducted by the National Academy of Sciences, could be valuable in promoting public education on all facets of the human predicament. To fill that role, the study must be supervised by scientists from the critical disciplines that can evaluate limits to the human enterprise—ecology, climatology, earth sciences, toxicology, epidemiology, and so on—not just demography and economics. The dependence of social and economic systems on the unimpaired functioning of ecosystems must be made explicit. Social sciences can perform a crucial role by designing economic and political systems that operate within the constraints set by physical, chemical, and biological systems. But both the need for this task and the understanding necessary for carrying it out are not yet appreciated fully by many social scientists. To put it starkly, ecology must take precedence over economics; nonsensical notions about perpetual growth of Gross National Product must be put to rest once and for all. Indeed, for the rich nations, economic growth is more the disease than the cure.

Some educational initiatives that could be taken in the United States, once the appropriate milieu is created, include:

1. Introduction of basic demography and the subject of population con-

trol in all high school curricula, in a context of civic responsibility. Every citizen should be at least as familiar with the demographic-environmental status of society as with its economic status.

2. Revitalization of the Council of Environmental Quality with increased staffing, to provide a steady flow of public information on the nation's environmental status. The Council's charge could and should be expanded to include global monitoring. Its reports should include timely information on changes in U.S. and world demography; in global climate and concentrations of greenhouse gases and other atmospheric pollutants; in acidification of lakes, streams, forests, and agricultural areas; in rates of soil erosion and groundwater pumping and recharge; in yields and compositions of fish harvests; and in energy use and efficiency in various sectors of the economy; and data on the preservation or loss of biodiversity. This kind of information should be generated and released to the public as regularly as economic information is today.

3. Coverage of these indicators of demographic, ecological, and climatic trends in public media, with the same level of attention as is now provided for stock market indexes, the prime interest rate, the trade deficit, leading economic indicators, and the value of the dollar. Ecological commentators should be as omnipresent as economic commentators are today (their predictions can hardly help being more accurate).

Heightened public awareness of population-related issues would open the way for instituting more substantive policies to deal with American and global population problems. A national campaign to lower the U.S. birthrate a little further—"patriotic Americans stop at two; real heroes have only one or none"—could hasten the day when a gradual population *decline* is initiated. The birthrate that has prevailed since about 1973 will eventually stop U.S. population growth (assuming net immigration is kept low) and lead to a very slow decline, but not before the U.S. population has surged beyond 300 million. With a slight further reduction in fertility now (from 1.9 to 1.6 children per couple, say), we could hasten the end of growth and the onset of a decline without exceeding 300 million.

Such a campaign should aim to increase public knowledge and practice of birth control, especially among young people. The costs to themselves and to society of children being born to and reared by single mothers, especially those still in their teens, should be clearly explained. The message that bearing and rearing children is an important responsibility, not to be taken lightly, should be imparted clearly early in life. Society has a stake in maximizing the well-being of and opportunities for future generations. It therefore has an interest in seeing that future parents are prepared for their responsibilities, including the ability to make decisions about childbearing—when to have each child, how many can be reasonably cared for, and so on—in the light of society's needs. And contraception—the means to carry out those decisions—should be available to all.

In our opinion, abortion should be available as a backup when contraceptives fail, as they sometimes do. Abortion has, unfortunately, become a highly politicized subject in the United States; people differ widely in their moral views about it. No sensible person views it as anything more positive than the lesser of evils in certain situations—and this circumstance adds to the difficulty of defending it morally. But in a world that is demonstrably overpopulated, where millions of people are, one way or another, denied access to contraceptive measures (if only because they are uninformed), where tens of thousands of children die daily from hunger and extreme poverty, and women clearly will seek abortion when they feel it is right (regardless of legality and even at the risk of their own lives), to outlaw abortion would be a criminal act against humanity.

In the United States of late, research and development of improved contraceptive methods and materials has been, sadly, neglected. In most other developed countries, a wider choice of methods is available and active programs are under way to develop new ones. The limited range of contraceptive choices in the United States, along with inadequate public education, helps inflate our disgracefully high abortion rate. Much more support for contraceptive research and development is needed, as well as some legal-liability relief for drug companies that develop and market them.

In conjunction with developing an overall U.S. population policy and improving our family-planning program, a review of immigration policy should be instituted. This time, the question of the future population size of the United States should be a major consideration. Few areas of public policy involve more, or more difficult, ethical choices. But our population growth is unlikely to end in the foreseeable future unless our policies address the imbalance between immigration and emigration. This reevaluation should include consideration of the role of U.S. foreign policies (including trade and investment policies) in generating immigration from poorer countries.

Finally, after clearly establishing a policy goal of population reduction for the United States, helping other nations control their population growth should return to a prominent place on the American foreign-policy agenda. In the 1970s, many developing nations opposed family-planning assistance from the United States, observing reasonably enough that Americans preached population control but did not practice it for themselves. Many saw such policies as a means for the rich to suppress the poor and grab more resources. Such concerns felt by developing nations are not unreasonable, although neither the rich nor the poor can afford population expansion as a way of controlling resources in this increasingly interdependent and fragile world.

Fortunately, opposition to aid for population programs has largely faded since 1980 as the U.S. birthrate remained below replacement level (which would eventually bring an end to growth) and as population-related problems in poor countries have become increasingly obvious. Indeed, the recent resurgence of population growth in the developing world is at least partly traceable to the Reagan administration's failure of leadership in population assistance. That leadership should now be reasserted; more, it needs to be augmented by new leadership in addressing and dealing with the emerging global environmental problems that threaten all of us, rich or poor.

The time is particularly propitious; the Soviet Union's experiments with *glasnost* and *perestroika,* and the relatively peaceful revolution in Eastern Europe have created an opportunity for the world's military superpowers to put aside the East–West confrontation, enlist the help of other economic superpowers (Japan and the European Economic Community), and begin to tackle jointly the most crucial problem of the twenty-first century: controlling human numbers and keeping the Earth habitable.

REFERENCES

Brown, Lester R., et al., *State of the World 1990.* New York: Norton, 1990.

Council on Environmental Quality and U.S. State Department, *The Global 2000 Report to the President.* Washington, DC: U.S. Government Printing Office, 1980.

Daly, Herman E., *Steady State Economics.* San Francisco: Freeman, 1977.

———, and John B. Cobb, Jr., *For the Common Good.* Boston: Beacon, 1990.

Djerassi, Carl, "The Bitter Pill," *Science* 245 (July 28, 1989), pp. 356–361.

Ehrlich, Anne H., and Paul R. Ehrlich, *Earth.* New York: Franklin Watts, 1987.

Ehrlich, Paul R., and Anne H. Ehrlich, *The Population Explosion.* New York: Simon & Schuster, 1990.

———, "World Population Crisis," *Bulletin of the Atomic Scientists* (April 1986), pp. 13–19.

Ehrlich, Paul R. and John P. Holdren, "Impact of Population Growth," *Science* 171 (March 26, 1971), pp. 1212–1217.

Hardin, Garrett, "Cassandra's Role in the Population Wrangle," in Paul R.

Ehrlich and John P. Holdren (eds.), *The Cassandra Conference: Resources and the Human Predicament.* College Station: Texas A&M University, 1988.

Holdren, John P., and Paul R. Ehrlich, "Human Population and the Global Environment," *American Scientist* 62 (May–June 1974), pp. 282–292.

Ornstein, Robert, and Paul R. Ehrlich, *New World/New Mind.* New York: Doubleday, 1989.

Pearce, Fred, "Methane: The Hidden Greenhouse Gas," *New Scientist* (May 6, 1989).

Schneider, Stephen H., *Global Warming: Entering the Greenhouse Century.* San Francisco: Sierra Club, 1989.

United States Commission on Population Growth and the American Future, *Population and the American Future* (7 vols.). Washington, DC: U.S. Government Printing Office, 1972.

Vitousek, Peter M., Paul R. Ehrlich, Anne H. Ehrlich, and Pamela Matson, "Human Appropriation of the Products of Photosynthesis," *BioScience* 36 (1986), pp. 368–373.

World Commission on Environment and Development (Gro Harlem Brundtland, Chair), *Our Common Future.* New York: Oxford, 1987.

SIR CRISPIN TICKELL

Environmental Refugees: The Human Impact of Global Climate Change

Sir Crispin Tickell has been the British permanent representative to the United Nations since 1987. He joined the British Diplomatic Service in 1954, and has had a widely varied diplomatic career. He was ambassador to Mexico from 1981 to 1983, a deputy under secretary of state for economic affairs in 1984, and in charge of the British overseas aid programme from 1984 to 1987.

As a long-standing environmentalist and student of climatology, Tickell is the author of *Climatic Change and World Affairs* (1977 and 1985), and has written and lectured widely on the social impact of global climate change.

I once attended an occasion at the Royal Institution at which the members, dressed in black tailcoats and starched white shirts with bow ties, awaited the arrival of the distinguished archaeologist Louis Leakey. As the clock struck nine piercing notes, Professor Leakey, dressed in an open-necked bush shirt, marched from doors behind the stage, mounted to the podium, and gazed at the eminent gathering before him. *"Animals,"* he said, *"let none of you forget that you are animals."*

It was a useful reminder. We have a long history, going back thousands of years, of thinking otherwise. From the earliest recorded days, people have placed themselves apart from and superior to other living things. Now we may be less sure of the frontier line. Beavers reconstruct the environment. Even birds shape and use tools. Yet we still congratulate ourselves on our souls, we still talk about conquering nature. A better way of putting it would be to say that we have been a very successful species within nature. We can adapt to the jungle, the desert, or the ice cap; we have changed the land surface to suit our needs; we have dealt with the worst diseases and raised average life expectancy; we have changed the genetic inheritance of plants and other animals for our food requirements; and we face no serious challenge from any other animal species. In short, as supreme predators we have enjoyed the luxurious opportunities of a species in rapid expansion.

That, of course, is the problem. I want to talk—as one animal to others at least as distinguished as those addressed by Professor Leakey—about an aspect of the predicament we have created for ourselves. The predicament is simply stated. First, like all successful species we have multiplied our numbers enormously; and second, we have affected greatly the environment of which we are a part. The two are now coming together as they so often do.

For biologists, a familiar experiment is that of the Petri dish. Petri dishes are round plates with transparent food on them disposed to allow an investigator to see colonies of microbes with the naked eye. From small beginnings the microbes multiply at an accelerating rate. They are at their most prolific as they reach the edge of the dish. Then the food runs out, the microbes die in their multibillions, and extinction takes place.

Now we are not microbes, and we do not live on Petri dishes. But if the microbes could reflect on their predicament, they would see the politics of their situation as a desperate rush for survival at the expense of all who stood in their way. There would be no time to talk about the nice balance between resources and population. For us, there is still a chance. The threat to our ways of life is visible, but more so in some places than in others. It is also clearer in abstract than in personal terms. My concern is how probable changes to the environment, particularly in global climate, could affect millions of our fellow human beings: how they live, where they live, whether they live. We face the specter of major human displacements.

It is scarcely necessary to speak of the multiplication of human numbers. In 1930 there were around 2 billion people in the world, and by 1975 that figure had doubled. Short of a catastrophe, there will be well over 6 billion people by the end of the century. Their distribution will not correspond to the distribution of the Earth's resources. Furthermore, the number of their accompanying domestic plants and animals represents a population explosion of its own. If some discriminating bug were to kill us all off tomorrow, the present forms of many of the plant and animal species we have so carefully selected would soon follow.

The changes we have made to the environment are sometimes too familiar to be recognized. But a visitor from 10,000 years ago would find a world transformed. The clearing of forests for agriculture, fuel, construction, and weapons of war (remember the wooden ships of Greece, Italy, and England); the cultivation and sometimes irrigation of vast areas of land; the spread of scrub and desert; and the building of towns and cities drawing their sustenance from all around them (more than 1 percent of the United States is now under concrete) have together changed the face of the Earth.

In the last 200 years change has been still more drastic under the impact of industrialization. That impact has been sharply different in different areas. In countries we label industrial, change grew out of previous history and was sustained by an agricultural revolution that increased food productivity greatly. The physical environment of these countries was resilient, with big natural resources including regular rainfall. The result was that although cities spread and the environment was much altered, little irremediable damage was done.

The same cannot be said for the rest of the world in all its variety. Change has not grown naturally out of the past: in many cases there was no preceding agricultural revolution to sustain an industrial society. Time was telescoped everywhere. Environment and resources were likewise different, with vulnerable and delicate ecosystems in forest and desert, poor leached soil easily destroyed, unreliable patterns of rainfall, all made worse by rapidly growing populations and often well-meaning misapplication of methods and technology useful elsewhere.

More recently we have become aware of what could be the trigger for the greatest changes of all: those in the chemistry of the air that have led to acid deposition in areas downwind of industry, to holes in the friendly shield of ozone that makes life possible, and to the prospect of global climate change.

Until now climate has usually been regarded, certainly by economists, as one of the invariable factors for all practical purposes affecting human society. In fact, climate has never been invariable. It is in constant change, but usually within well-behaved limits. Substantial change has rarely been identifiable within a human timescale. But people have of course felt and remarked on it.

Those living on the margins of the four geographical zones of ice, temperate areas, desert, and tropics have long been sensitive to small changes. When the little ice age began in the fourteenth century, those growing wheat in Greenland or Iceland lost their livelihood; when the warming of the present century began, mountain glaciers shrank and new land was opened for cultivation.

But the idea that mankind could affect the climate is relatively new. Somehow the littleness of man and the vastness of the sky made it an improbable proposition, and to many it still is. But evidence the other way is strong, and it is accumulating fast. The relationship between land use and climate, and in particular between trees and rainfall, has been demonstrated. More recently we have established—it is a fact and not a hypothesis—the relationship between the quantity of carbon dioxide in the atmosphere and average global temperatures. This is not the occasion for a disquisition on the greenhouse effect or, perhaps more accurately stated, the global heat trap. Without such a trap, life as we know it on the Earth's surface would be impossible. The Earth would be like Mars, from which too much heat has escaped, or Venus, within which too much heat is retained. We have become used to the present broad equilibrium of heat entry and heat loss, and our pattern of existence relies on it. But by increasing the volume of certain trace gases in the atmosphere, we are likely to move the limits of the equilibrium in ways that cannot so far be determined.

We know that levels of atmospheric carbon dioxide and methane are increasing each year. The other principal greenhouse gases are chlorofluorocarbons, which were unknown before the 1950s, and nitrogen oxides.

Most of the increased carbon dioxide comes from consumption of fossil fuels, and the rest from the burning of forests; methane comes from land clearance, natural-gas release, fields such as rice paddies, biomass burning, and ruminants; the chlorofluorocarbons come from industry (refrigerants, aerosols); and most of the nitrogen oxides comes from land conversion, fertilizers, and fossil-fuel consumption.

The effects, now and in the future, of the increase in these gases remain a matter of dispute; but the conventional—and, I add, conservative—wisdom, with some heretics on each flank, is that global mean temperature will rise by between 1° and 2°C by 2030, and by more than another 0.5°C by 2050. It is unlikely to rise by less. It could rise by much more. Already the mean temperature of the Earth has risen by 0.5°C over the last hundred years.

This range of increases may seem small. But so was the drop in global mean temperature of around 4°C which prevailed during the last ice age. Moreover, the use of an average figure conceals the drastic changes that would take place at certain latitudes. Whereas the change around the equator might be very small, in temperate areas it would be substantial and at the poles could be a 7° to 10°C increase. Efforts have been made to estimate the likely effects for particular latitudes and regions. But models so far are inadequate, and we have no more than general ideas to work with. If we take into account what happened in previous periods of warming, there would be a shift of temperate conditions northward in the Northern Hemisphere and southward in the Southern Hemisphere; there would be a more lively heat exchange between the equator and the poles, creating greater instability between them with more storms, droughts, and deluges; and there would be a rise in mean sea levels from a combination of thermal expansion and melting ice.

As the oceans represent an element of inertia in the global thermostat, the rise in sea levels would probably take longer than the rise in atmospheric global temperatures. At present the best guess is a rise of between 24 and 34 centimeters in the next sixty years, but it could be less or more. The current rate of rise is around 2.4 millimeters a year.

Changes of this kind have been known before in the Earth's history. Living things made the necessary accommodations with some loss of plant and animal species. For us the problem is less change itself than its speed. Change is at present taking place at a rate some ten times faster than the average over the past 10,000 years, and at a rate many times faster than that since the last ice age. Indeed the rate of change could be so fast it could cause disruption to ecosystems comparable to those which caused major extinctions of species in the past. A change in rainfall patterns of only 300 to 500 kilometers north or south would lead to a different covering of the Earth. Some plant species might be able to move and adapt themselves to new soils, but other, longer-lived species such as trees might be unable. Animals—from insects to fish to elephants—would face major problems of adjustment. For all there would be many losses and some gains, and the pace of evolution might quicken to fill the gaps. The likely rise in sea levels would flood many existing coastal areas, change existing river systems and aquifers, and possibly alter the pattern of ocean currents and the distribution of marine resources.

No animal would be more affected than we. During previous periods of warming or cooling, our species was able to respond with its feet. The last period of cooling showed a human invasion of the Americas and Australia, over the land bridges that emerged when sea levels fell. The last period of warming showed a human invasion into the areas liberated from the ice: a good example is the United Kingdom itself. But in a new and more drastic period of warming, there would be few places for people to go to. Other people are there already. We have left ourselves no room to maneuver. Warming might release land for settlement in what is now Arctic tundra, but there is no imaginable way in which populations living elsewhere in areas of sudden environmental stress could pick up their bags and move. The barriers are up, and for good reasons or bad, most people already find it harder than ever to leave their place of birth.

Yet already the number of the world's refugees is steadily increasing. Accordingly to a present definition, a refugee is one who, "owing to well-founded fear of being persecuted for reasons of race, religion, nationality, membership of a particular social group or political opinion, is outside the country of his nationality and is unable or, owing to such fear, is unwilling to avail himself of the protection of that country." There were fewer than 5 million refugees in 1978 and almost 14 million in 1988. Most of them were fleeing from war, persecution, and the other consequences of political breakdown. In camps or on the move were almost 6 million Afghans, almost 2.5 million Palestinians, over 1 million Mozambicans, over 1 million Ethiopians, and half a million Iraqis. Less in number, but just as big in terms of local impact, were 400,000 Somalis, 400,000 Angolans, 335,000 Cambodians, 300,000 Sudanese, and 150,000 Central Americans. These figures do not include the millions of undocumented aliens, asylum seekers, and those simply looking for a better way of life—so called economic migrants—in the United States, Europe, and the Middle East. Nor do they include those displaced within and outside their own countries by famine, drought, flooding, or other environmental hazards (a number estimated at well over 10 million people).

It requires a leap of the imagination to work out the numbers that would be on the move in the event of global warming on present estimates. A heavy concentration of people is situated at present in low-lying coastal areas or along the world's great river systems. Nearly one-third of humanity lives within 60 kilometers of a coastline. A rise in mean sea level of only 25 centimeters would have substantial effects. The industrial countries might be able to construct new sea defenses to protect vulnerable areas, but even they would have difficulty in coping with high tides and storm surges of a kind that might be more common.

For most poor countries such defenses would be out of the question. Many of those living and working in, for example, the delta areas of the Nile, the Ganges, and the Yangtse would be forced out of their homes and livelihood. Some islands, such as the Maldives in the Indian Ocean, and Kiribati, Tuvalu, and the Marshall Islands in the Pacific, would soon become uninhabitable. Bangladesh, with its population of more than 100 million, and Egypt, with its population of around 60 million, would be particularly affected. A further rise in sea levels up to half a meter and beyond would have more drastic results. The world would look a different place.

For those living well above sea level, global warming would have other disruptive effects with differences according to latitude. In broad terms, the tropics would be least affected; but those dependent on such annual events as the monsoon or summer rains might find new irregularities in the weather system. Recent droughts in the Sahel have illustrated the vulnerability of people living in or on the fringes of arid zones, and current population levels would almost certainly prove unsustainable. Countries in which economic resources, including fresh water, are already strained by increasing human population would suffer most.

The present temperate areas, where most of the world's industry and agricultural production now lie, would not escape. The food surpluses that at present act as a buffer stock to cope with deficits elsewhere could disappear quickly. The climatic variation would probably be greater in these countries than toward the equator, and such areas as the northern Mediterranean, the American Midwest, and the southern parts of the Soviet Union would be particularly vulnerable. But there might eventually be compensation in opening up land farther north, and ways of mitigating change through application of technology might be found.

What, then, could be the scale of the human problem thirty, forty, or fifty years from now? Even allowing for piecemeal, gradual responses year by year to the imperatives of change, it would be very large. So long as the problem remains unsolved in the short term, so it steadily accumulates in the long term. Plucking a figure from the air, if only 1 percent (a very low estimate) of a future world population of 6 billion were affected, that would still mean some 60 million migrants or environmental refugees; and 5 percent (again a low estimate) would produce 300 million. Even 60 million would represent a problem of a magnitude no one has ever had to face.

Nor is it all. Refugees create their own problems. Under its present mandate the United Nations High Commission for Refugees provides protection and assistance for refugees and does likewise for others when mandated by the UN General Assembly. In some cases refugees can return to their country of origin. In others they can be resettled. But economic migrants or environmental refugees fall into another category whether they stay in their own country or cross into another.

At present they are a phenomenon mostly of poor countries. Richer ones try to keep them out. Shelter, food, and medical care are hard to find. There is little prospect of return. They often come from another environment (for example, highland Ethiopians are forced down to the plains), and they bring with them alien customs, alien religious practices, alien eating habits, alien agricultural methods, and—not least—diseases with susceptibility to local pathogens to which they have built no resistance. Most have great difficulty in adjusting themselves to new circumstances. Like normal refugees, they depend on charity. Resettlement is never easy, and full assimilation is rare. In any numbers they tend to spread their poverty around them and to compound the problem from which they first tried to escape. In a warming world, refugees would constitute only one of the myriad animal species trying to cope with the disruption of a way of life; and as with other creatures, death rates among them inevitably would be high.

This bleak picture brings out a dilemma well expressed by the prominent scientist A. V. Hill in 1952. He then wrote in *Nature* magazine:

Some might [take] the purely biological view that if men will breed like rabbits they must be allowed to die like rabbits. . . . Most people will say no. But suppose it were certain now that the pressure of increasing population, uncontrolled by disease, would lead not only to widespread exhaustion of the soil and of other capital resources but also to continuing and increasing international tension and disorder, making it hard for civilization to continue: would the majority of humane and reasonable people then change their minds? If ethical principles deny our right to do evil in order that good may come, are we justified in doing good when the foreseeable consequence is evil?

Perhaps no one will wish to answer that question, but it still needs to be asked. Our instincts of compassion may move us one way, and our sense of self-preservation and preservation of our families, in another. It must be our aim to recognize but avoid that painful dilemma.

In virtually all countries the growing numbers of refugees would cast a dark and lengthening shadow. Within a country they would represent a dangerous factor in the certain and growing difficulties of social and economic management. We are familiar with the strains that famine, drought, flooding, and other disasters can bring to governments. Some can cope, some evidently cannot. But few outside the industrial world have the structure or resources to manage a continuing crisis. Secondary effects of disorder, terrorism, civil war, economic breakdown, or even bankruptcy could become endemic. We are all too familiar with them already. Witness the lingering agony of Lebanon, or the slow slide of Peru into chaos. In such circumstances the care and management, let alone resettlement, of local refugees would be a challenge many governments could not hope to meet.

Among countries and regions there would be still larger difficulties. To a greater or lesser extent all would be suffering and undergoing adjustment. Thus willingness to help others would be limited, the more so if it threatened to put at risk the adjustment process at home. Some of that process might lie in setting land aside for fallow or reforestation. There are already little patches in Africa and India protected for this purpose—oases of green in brown landscapes—but the condition of success is the continued absence of man and his domestic animals. When productive land and fresh water are precious, no one welcomes intruders. There are no more ancient cause of conflict between peoples than land and water. In time of trouble the pressure of recognizable aliens is liable to ignite popular resentment with the speed of a brushfire; we have seen recent violent incidents of this kind in Mauritania and Senegal.

In industrial countries, many feel—rightly or wrongly—that there is an absolute limit to the number of people from other countries and cultures who can be absorbed without damaging social cohesion and national identity. With substantial unemployment, primary immigration into Western Europe and the United States has diminished to a trickle over the last fifteen years, and resistance to immigration has become popular politics. Any aggravation of the refugee problem would only strengthen such resistance.

But even if some people and governments wished to seal themselves off from the rest of the world, they could not do so. In no country or city can the rich fortify themselves for long against the poor. All form part of an increasingly interdependent economy. Land frontiers can always be penetrated. The northward movement of Mexicans and other Latin Americans across the long southern frontier of the United States has so far proved irresistible. Every year parts of the United States develop more Hispanic characteristics. Nor are short sea crossings a real barrier. Desperation could push Africans into Europe, Chinese into the relatively empty parts of the Soviet Union, and Indonesians into northern Australia. Sheer numbers could swamp most efforts at control.

We forget, at our peril, that civilization is a fragile thing. The first cities appeared only about 5,000 years ago and until recently were so unhealthy their populations required constant replenishment from the surrounding countryside.

As John Reader recently wrote in *Man on Earth:*

In the brief space of time that civilization has been a feature of human existence, it has not demonstrated any tendency to produce a well-regulated steady state whereby people are well fed and secure, generation after generation. Civilization is distinguished more by its erratic cycles. . . . Time and again it has risen dramatically from the field of human endeavour, then collapsed and fallen. Human ingenuity drives the process. Human inability to impose adequate restraint brings it down. Inventions provide the initial impetus, intellect supplies methods of application and solutions to problems that arise as the system swells and grows, but in every instance so far, the uncontrolled growth of civilization has ultimately thrown up more problems than human intellect can solve.

The key phrase here is "inability to impose adequate restraint." For we are not helpless before this problem. It is one we have created for ourselves. Industrial society is a highly artificial construct with an internal logic not unlike one of its own machines. It rests on the use of a particular kind of energy stored since the early history of the Earth—and has led to a transformation of the land and generated a vast increase in our species, thereby destroying many others. The influence of industrial society has spread everywhere, and its success has created boundless expectations. Now we are getting the bill.

We may not be able to meet it. But we can do a lot if we have a mind to. There is not much time. It is too late to think of prevention. The chemistry of the atmosphere has already changed, and even if we could cut back manmade emissions of greenhouse gases immediately, the effects would be with us for hundreds of years. But we can certainly mitigate the problem.

This is not the occasion to set out a program of international action. But such a program would include the elaboration of new energy policies designed to conserve energy (good examples range from better insulation materials to new "miracle" light bulbs); reduce consumption of fossil fuels, especially coal and oil; develop alternative sources, from nuclear to solar; and reduce burning of fuelwood in poor countries. It would include a reorientation of industrial policy to avoid use of chlorofluorocarbons and curtail noxious emissions. And it would include changes in land use to promote reforestation and wood harvesting; new methods of agriculture, including agroforestry; and the management of wildernesses, deserts, and human landfills through application of old and new techniques, including biotechnology.

Apart from mitigating effects, we must also accept the need to adapt. With greater knowledge of what is likely to happen, we can prepare ourselves for change. This may require small alterations as well as big ones: from changes in diet (certain plants will do better than others in a carbon dioxide–rich world) to major public investment in energy, desalinization to obtain fresh water, and protection against rising sea levels. To cope with the speed of likely change, nature may well need a helping hand; for example, in transferring trees from one area to another. We need to do more through a variety of methods to curb human population increase, to inculcate more understanding of the environment and other living creatures within it, and to support worldwide development only on a basis sustainable over generations. This means shifting emphasis from simple exploitation of resources to better use of them, and more respect for the people involved. To mention but one instance: In about forty projects approved for financing by the World Bank in agriculture and hydroelectricity between 1979 and 1985, provision was made for the relocation, voluntary or not, of over 600,000 people in twenty-seven countries. Was this really necessary or justified?

All efforts to cope with these problems must be as global as the problems themselves. But very little can be done without a wider understanding and measure of consensus. Perhaps the first successful step was the Stockholm Conference on the Environment held in 1972. In the following years there were many cries of alarm—most went unheeded. Institutions and conferences multiplied but without much effect. With the publication of the report of the Brundtland Commission on Environment and Development in 1986, there has been a sea change in public attitudes. The most practical results so far have been the Montreal Protocol of 1987, a product of the Vienna Convention of 1985, to limit production and use of chlorofluorocarbons. "Greenhouse effect" has become a household phrase, and the idea, if not the implications, is better understood. There was even a debate in the United Nations on the subject in autumn 1988, when I made my first speech on that subject to the General Assembly. It was followed by a useful and practical resolution.

It fell to me in New York on May 8, following the seminar held by the British Prime Minister in London on April 26, to set out the main elements in the British government's approach to the international aspects of the global warming problem. In a few words, these elements are the formation of an umbrella convention to set out general principles with detailed protocols on specific aspects as they arise; the strengthening and development of existing UN institutions, including the United Nations Environment Programme, the Intergovernmental Panel on Climate Change (a body required to set out the science of the problem), and even possibly the Security Council under Article 34 of the UN Charter; and acceptance of the fact that global warming has arisen essentially from the process of industralization. It follows that industrial countries must give leadership and help to the rest of the world in managing the three linked problems of energy, industrial change, and land use. I wish it could be said that international action on these lines could provide solutions to the various problems, including the dilemma of environmental refugees. But I do not believe that it could, even if it were able to mitigate such problems. So far, few have even considered the refugee aspect—except perhaps in a few chilling throwaway lines.

There is no accepted definition of an economic migrant or an environmental refugee. Of course, these refugees have been with us for a long time even if they have not always been labeled as such. The shantytowns of poor countries are full of them. The United States had many of its own after drought in the Midwest in the 1930s.

I do not have magic solutions to suggest for still greater problems that may lie ahead. In the long run, there is no alternative but restoring equilibrium between resources and population, and achieving flexibility through optimum rather than maximum population density. It has already been done in one or two places. Eventually, we are all dead, including countless refugees. It is today that matters. The first step to wisdom is to recognize the problem. The next is to do all possible to prepare for it. As the movement of refugees across frontiers would be extremely unwelcome, and could not be resisted even by force, governments must work to manage the problems themselves. In doing so they would need, and have a right to, help from the international community, particularly from those in it who have contributed most to the creation of global climate change. Action to accommodate refugees across frontiers would be immensely more difficult, and without international agreement could risk creating tension, disorder, and conflict on a major scale.

Science is full of surprises. Things do not necessarily change in linear or step-by-step fashion. A measure of unpredictability and chaos is endemic. The beating of a butterfly's wings in Manhattan may contribute to an eventual deluge on the other side of the Atlantic. But even if we allow for surprises and chaos, it appears in 1990 that the world is warming up at a rate unknown in historical times, and that with the present increase and distribution of the world's population, we face disruption of existing society with distressing consequences for millions of displaced human beings. As animals, we are both tough and adaptable. But our toughness and adaptability might, for many, be tested beyond endurance. We have grown, lived, and flourished as elements in specific natural surroundings. Those surroundings or ecosystems could be so damaged that they may fall apart, as they often have in Earth's history, to be replaced by different ones. We must look to our present companions in life in their marvelous complexity, as well to ourselves, if we and they are to survive and prosper. We want to avoid A. V. Hill's cruel dilemma. We want to avoid the politics of the Petri dish. Whether we can I do not know. We should at least try.

NOEL J. BROWN

Global Warming and Climate Change: Challenge in Search of a Strategy

Noel J. Brown is a special representative of the executive director, and regional director of the United Nations Environment Programme North America. A citizen of Jamaica, he has lectured extensively in the United States and the Caribbean. His career includes extended service with the United Nations, where he was formerly political affairs officer in the Department of Political and Security Council Affairs. He has also headed the North American Office of the United Nations Environment Programme.

Dr. Brown has represented the United Nations Environment Programme at many important conferences, including: the UN Conference on Habitat and Human Settlements (Vancouver, 1976), the UN Conference on Law of the Sea (Geneva, 1980), and the Conference of Plenipotentiaries on the Protocol of Chlorofluorocarbons to the Vienna Convention for the Protection of the Ozone Layer (Montreal, 1987).

ADDRESSING THE PROBLEMS

We are hopeful that the initiatives of the Greenhouse/Glasnost symposium will receive a positive response and provide a clear signal to the world that even the most powerful political and military adversaries, the Soviet Union and the United States, could become ecological allies in the service of the Earth. And this would seem natural since they are the only two countries possessing "a strategic capability"—that is, the ability to change the status quo by force. Perhaps they will cooperate to use their special advantages to change the *environmental* status quo. In any event, the United Nations Environmental Programme (UNEP) observed the proceedings of the symposium and approved of them.

But UNEP tempered the approval with this: "Whereas the leadership of the Soviet Union and the United States on this critical challenge is to be commended and actively promoted, we cannot afford to lose sight of the global nature of the climate change issue and the importance of global cooperation." Perhaps Bill Clark captured this global dimension more precisely in his very thoughtful essay, "Managing Planet Earth." He wrote, "It is as a global species that we are transforming a planet. It is only as a global species pooling our knowledge and coordinating our actions and sharing what this planet has to offer that we could have any prospect of managing the planet's transformation." (*Scientific American*, September 1989, p. 47)

Nowhere is this more ominous than on the question of global climate change—and nowhere is the need for broad global cooperation more imperative. Humanity confronts a challenge of unprecedented scope and magnitude that will undoubtedly affect the way we live and work; how we farm and what we eat; our modes of production, consumption, and transportation; our recreation and leisure, health and well-being; even how we raise our children. Despite "considerable uncertainties," there appears to be enough consistency in available information to warrant at least a policy of prudence and the formulation of broad cooperative strategies. There literally is no time to waste. Without wishing to be alarmist, but based on the considerable data at its disposal, UNEP has concluded that the decade of the '90s—4000 days—may be our last window of opportunity to adopt the necessary measures to stabilize the Earth's climate system, or at least to slow the rate of change.

As my executive director stated in his World Environment Day message last June 5th, "We shall win or lose the struggle [for our environment] in the first years of the '90s. The issue is as urgent as that." And yet there is no generalized sense of urgency—not because of any absence of warning, repeated with ominous frequency, with which we are by now all too familiar.

For years, scientists have cautioned that if the present trends continue of increased fossil fuel use, deforestation, and manufacture of CFCs, build-up of greenhouse gases is likely to alter global temperatures by between 1° to 4° C in the next thirty to fifty years, with consequences that are as unpredictable as they may be unmanageable. Compounding these difficulties is the fact that there is, as yet, no clear understanding of specific regional effects—only broad symptoms and speculations. Let me describe a few.

The energy retentiveness of the atmosphere will increase because of added loads of greenhouse gases. Wind patterns will change with probable increases in hurricane frequency and tornados. The killer storms of 1988 and 1989, Hurricanes Gilbert and Hugo, may be just a mild prelude of ominous things to come.

Wind shear will become more dramatic and potentially devastating. Aridization and dust from windblown erosion is likely to increase. This will damage natural habitats and threaten human communities, even populations.

Already the world faces a loss of about 26 billion tons of productive soil a year. This is worsened by desertification of lands where the toll in losses of productive soil capital equals an area of 11 million hectares or about 27 million acres a year. Nearly one-third of the world's arable land will be destroyed in the next twenty years from existing land-use patterns and erosion.

Nearly 44 percent of the land resources of Africa, and 43 percent of those in South Asia, are subject to drought. Moreover, 47 percent of soil resources in South America and 59 percent in Southeast Asia are prone to nutritional deficiencies or suffer the impact of toxic substances. Barely 15 to 18 percent of the soils in South America, Africa, and Asia may be said to have no serious limitation for agricultural use. That is why soil conservation and environmentally sound land-use patterns have been elevated to the status of a priority item on UNEP's new planetary agenda. Effective land management, however, must take into account the new and unpredictable risks of windblown erosion.

We must also worry about the possible interruption of one of nature's most efficient fail-safe systems, forests. As a result of massive and accelerating deforestation, we are destroying one of the principal sinks for carbon dioxide, and in the process consume our life-giving source of oxygen regeneration.

Trees, as you know, absorb carbon dioxide, store it in the form of organic compounds, and release oxygen. Because of their sheer height and mass, trees are, by far, the most efficient laboratory for photosynthesis operations. Replacement of forests by any other form of vegetation with a smaller biomass-per-unit area would probably add significant amounts of carbon dioxide to the atmosphere. Replacement of the entire Amazon forest by grass, for example, would probably release an amount of carbon dioxide into the atmosphere equal to about 20 percent of that of the whole world. Even assuming about half of the gas would be dissolved by buffers, namely oceans, rivers, lakes, and so on, there would still be a significant increase. According to some estimates, the effects of deforestation will result in adding between 1 and 2.5 billion tons of CO_2 to the atmosphere, thereby transforming forests from *sinks* to *sources* of carbon effluents—yet deforestation continues.

Forty percent of the world's tropical rain forests and tropical deciduous forests have disappeared already, and the remainder is defiled at the rate of 110,000 square kilometers a year or 20 hectares a minute—that is, one football field per second! The destruction has to be dramatically slowed and massive reforestation initiated as a matter of global policy. This is an eminently doable option—and a UNEP priority.

Another presumed effect of global warming is a rise in sea level from melting polar ice caps. This will cause considerable coastal inundation and a significant interruption in both coastal dynamics and coastal economics. Industries—tourism, harbors, ports, fishing—could very well be destroyed. UNEP, for example, estimates that a one meter rise in sea level could displace up to 15 million people in countries like Egypt and Bangladesh.

Lastly, as the land loses its biomass cover, it becomes much less moisture retentive. Cloud formation will be affected, and water vapor will shift to those places where the conditions are more appropriate.

Most plants are climate-oriented, following a long process of adaptation over geological time. Abrupt changes in rainfall patterns and local climate will alter the plant's intimate relationship to its environment. Unless the species is very hardy or climate-indifferent, this could create havoc with the world government's agricultural picture of the future. Perhaps it is timely to ask whether genetic engineers are experimenting with "transitional species" that will be able to withstand global warming and at the same time ensure future agricultural needs.

THE CALL FOR ACTION

The Global Conference on the Changing Atmosphere: Implications for Global Security, held at Toronto in June 1988, was convened to reinforce not only the warnings from scientists but to ask governments, the United Nations, nongovernmental organizations, industry, educational institutions, and the individual—in a word, all of us—to counter the degradation of the atmosphere. This international meeting was perhaps the most broadly based and authoritative of its kind. Foremost among the actions proposed were reduction of carbon dioxide emissions and other greenhouse gases, improvements in energy efficiency, and the halt of deforestation—the same appeals made at Sundance. The conference also called for a comprehensive global convention as a framework for protocols on the protection of the atmosphere.

Humanity has an agenda for managing its atmospheric resources. But does humanity have the will? And will it have the time?

Perhaps, and perhaps not. But at least humanity has at least one *certain* asset for the common good at its disposal. The United Nations, an established and seasoned global framework, has demonstrated its capability for global consensus-building, cooperation, and problem solving in such diverse areas as atmospheric ozone depletion, control of toxic chemicals in the export trade, and regulation of the transfrontier movement of hazardous wastes. The challenge, however, is how the world's peoples will use the United Nations—for creative and cooperative ends as well as competitive ones—in shaping the management of global change and stabilization of world climate systems.

Indeed, at the 43rd General Assembly last year, the representative of Malta submitted an item initially entitled, "Protection of Climate as a Common Heritage of Mankind," although it was subsequently amended to, "Protection of Climate for Present and Future Generations." To some, this would simply be another in the interminable United Nations agenda of issues reduced to little more than a debate—a shortsighted view, no doubt. The Malta initiative has placed climate change and the greenhouse effect on the global agenda, where it is now the subject of annual review and report, and where it will generate a new vocabulary of concern for the United Nations, as did general environmental issues some years ago, and sustainable development of nations only last year.

United Nations involvement is crucial to the developing countries now able to participate fully and continuously in an on-going global dialogue— and to shape an outcome consonant with their own interests. The result forms a kind of "global learning process," and a stimulation of global "consensus-building" that is likely to accelerate over time.

Moreover, the United Nations provides an objective forum where issues of equity can be addressed as the process of negotiating a new global bargain begins. The General Assembly emphasized the scope and magnitude of the need for global cooperation in its first decision: "The problem of climate change affects humanity as a whole and can only be confronted within a global framework so as to take into account the vital interest of all mankind." It went on to urge governments, intergovernmental organizations, nongovernmental organizations, and scientific institutions "to treat the problem of climate change as a 'priority matter'; " to put in place specific action-oriented programs and research, with specific time frames, on climate change that include a focus on "regional respect"; and to contribute human and financial resources to multilateral efforts to protect the environment from potentially devastating damage.

Recognizing the need for heightened public awareness and understanding, the General Assembly went on to encourage the convening of conferences on climate change and global warming at local, national, and global levels to make all government organizations—and the international community—better aware of the "importance and urgency of dealing effectively with all aspects of climate change resulting from certain human activities."

In seeking to lay the groundwork for a policy agenda, it called on the World Meteorological Organization (WMO) and UNEP, together with the Intergovernmental Panel on Climate Change, to initiate immediate action and lead, over the next eighteen months, a comprehensive review that will issue recommendations with respect to these five specific areas of inquiry:

1. The state of knowledge about the science of climate, with special emphasis on potential climatic change directly related to a global warming.
2. The study of socioeconomic impacts.
3. The formulation of possible policy responses by governments and others that may delay, limit, or mitigate the impact of adverse climate change.
4. The negotiation of relevant treaties and other legal instruments dealing with climate.
5. Elements for possible inclusion in future international conventions on climate.

These recommendations undoubtedly will strengthen the work of the Inter-Governmental Panel, created approximately one year ago to form a strategy for action, and will reinforce and give a new authority to efforts by UNEP in designing and develcping its bioregional strategy for the management of climate change.

UNEP welcomes the General Assembly's formation of this important task force as a major boost and validation of its own efforts.

UNEP'S CONTRIBUTION

Following the 1972 Stockholm Conference on the environment, the United Nations Environment Programme initiated an ecosystem strategy for the management of ocean space called the Regional Seas Programme. The program is now applied to nearly a dozen regional or semienclosed seas, involving more than 120 countries who share these coastal commons. It is premised on the fact that over a quarter of the world's population live in coastal areas, and over 90 percent of the world's currently exploitable resources are found in coastal waters. It is also premised on the fact that a nation's people are more likely to answer a call for cooperation when they must defend national interests in a shared dilemma or dispute than when they are called upon to support more generalized global causes.

So far, this strategy is implemented in:

1. the Red Sea and Gulf of Aden,
2. the Persian and Arabian Gulf,
3. the East Asian region,
4. the Southeast Pacific,
5. the Southwest Pacific,
6. West Africa, including the Gulf of Guinea,
7. East Africa,
8. the Southwest Atlantic,
9. the Caribbean,
10. the Mediterranean.

Through the Regional Seas Programme, UNEP has evolved a large global network of arrangements and principles supportive of marine coastal cooperation. Perhaps the most dramatic achievement of the program so far has been the Mediterranean, a dying sea. Covering less than one percent of the oceans' surface, it contained more than 50 percent of all oil tar ball pollution. The Mediterranean has witnessed the rise and fall of more civilizations than any other region in the world—and concurrently experienced more ecological stress than almost any other region. It remains the center of many ancient and recent antagonisms, including the Arab-Israeli conflict.

Hardly a place for cooperation, and yet the region's leaders seem to have succeeded environmentally where other statesmen have not. A high level of environmental cooperation exists under the Mediterranean Action Plan, with the result that the Mediterranean now shows definite signs of environmental improvement.

Based on this experience, UNEP has begun to direct its regional seas network to address the problems related to climate change and global warming. UNEP may yet provide the largest single network mobilizing developing countries in a systematic attempt to initiate action in *anticipation* of projected climate changes.

In 1987, "task teams" on the implications of climate change were established for six regions covered by the UNEP Regional Seas Programme: the Mediterranean, Caribbean, South Pacific, East Asian Seas, South Asian Seas, and the Southeast Pacific. In 1989, East and West African Seas were included. Two more, the Red Sea and Black Sea, are on the drawing board. These teams, using a common format developed at Villach, had a fivefold task:

1. To examine the possible effects of sea level changes on coastal ecosystems.
2. To examine the possible effects of temperature elevations on terrestrial aquatic ecosystems, targeting economically important species.
3. To examine the possible effects of climatic, physiographic ecological changes on socioeconomic activities.
4. To determine the various ecosystems that appear to be most vulnerable.
5. To prepare regional reports, assessing the general and specific problems of each region.

Although much work remains to be done, preliminary findings in the Caribbean offer some interesting insights on the process at work, as well as implications for policymakers. Topping the list are prospects of serious ecosystem stress and erosion of the deltas of the region's four major rivers, the Mississippi (USA), Rio Grande (Mexico/USA), Magdelena (Columbia), and Orinoco (Venezuela).

Other probable specific effects likely in the Caribbean targeted for further study are:

1. Possible increases in "storm activity." The Jamaican experience of 1988 and the Hurricane Gilbert syndrome signal more frequent, ferocious storms that will accelerate coastal erosion;

2. Loss of some wetlands;
3. Increased flooding of some coastal plains;
4. Salt intrusion and its effect on the world's freshwater supply;
5. "Hot snaps" affecting fishery resources;
6. Impacts on tourism in coastal areas.

While these six effects are highly provisional and are proposed cautiously, the task team was firm in its recommendations on several areas of policy: improve communication and information exchange; reduce uncertainties about regional impacts of a 1.5 degree per 20 centimeter global warming scenario by further employing enhanced data-generated case studies and modelling that will yield best-guess estimates of sea level rises and other climate changes; and continue the task team's interdisciplinary interaction to provide qualitative information to member states.

Findings of studies examining the effects of global warming on the coastal areas and seas in the East African and West African marine environment are being evaluated now.

But the case of the Mediterranean still seems the most dramatic and highly instructive. As a result of preliminary reports from the Mediterranean task team, already one government—Egypt's—has begun to examine the implications of sea level rise, and how land-use planning and population concentrations will have to be adjusted accordingly. The reason should be clear. Egypt's population now resides on only 4 percent of its land; 15 million of its 49 million people live within 30 kilometers, or 19 miles, of the Mediterranean coast.

Egypt convened, in cooperation with UNEP and the United States Environmental Protection Agency, "A National Crisis Conference on Global Warming and Climate Change" based on the scenario of a 100 centimeter, or 40 inch, rise in sea level within half a century. Already, Egypt's Mediterranean shoreline is eroded at the rate of 30 meters, or 100 feet, and more a year in some places. Based on these calculations, Egypt contemplates a 12 to 15 percent disruption of agricultural productivity and a 15 percent loss of Gross National Product.

Against this background is a growing fear that as much as one-fifth of Egypt's farmland will be affected as the sea level rises, and as many as 10 million people may become environmental refugees. The government of Egypt is intensifying studies on land-use patterns, coastal erosion, land reclamation, and wetland preservation, along with long-term development plans incorporating the possibility of unpredictable and changing conditions. Problems related to fresh water, nutrient discharge, and land subsidence are of priority consideration. Perhaps the Egyptian effort will create a precedent for orderly adaptation during the next century. But Egypt is not alone.

Assuming current models and programs are accurate, regional adaptations are not significantly implemented, and greenhouse gas emissions continue at present levels, then as many as 1 billion people residing near the world's coastlines could be affected by rising sea levels, creating approximately 300 million environmental refugees. Low-lying developing countries such as Bangladesh, Indonesia, and the Maldive Islands would be hardest hit. The government of the Maldives plans to host a conference for small island states threatened by sea level rise.

A UNEP-sponsored conference held recently in the Pacific Marshall Islands recommended that strategies should be readied to relocate up to 200,000 people in preparation for rising water levels because of a global warming.

FOR THE FUTURE

These initiatives are suggestive to me of a breakthrough in interest from developing countries. They must actively participate in the formulation of response strategies to predicted climate change. UNEP plays a fundamental role in placing critical environmental issues on national and regional agendas, and legitimizing the participation of governments that otherwise might feel this is a problem for the developed countries to solve. One can only hope similar initiatives will receive broad global support, particularly from the major industrialized states.

UNEP's efforts could be pivotal to the way developing countries approach negotiations at the global convention on climate change expected to begin once the Intergovernmental Panel concludes its work. Developing countries will have a fuller scientific grasp of the complexities of the problem and its possible effects on them, a better sense of available policy options, as well as an understanding of the specific agenda of the convention. They will bring a well-informed intelligence to the negotiating table formed by serious scientific investigations, and a working relationship at home between scientists and policymakers.

The Regional Seas Project launched by UNEP represents, then, a major part of the preparatory process leading up to negotiation, stimulating global learning and the habit of multinational cooperation in addressing global climate change—a problem of great complexity steeped in uncertainty.

The major industrial powers could facilitate this process of consensus-building and cooperation by first identifying what resources, technology, and know-how they are prepared to make available to assist developing countries during the adjustment period. This is being done now to control the rate of ozone depletion. Certain developed countries are cooperating with UNEP in a series of case studies to determine the costs involved in implementing and adhering to the Montreal Protocol among developing countries.

A world climate fund is advocated by UNEP. Some countries, such as Norway, have offered to contribute some $100 million every year; the Netherlands has offered 250 million Dutch guilders. In the words of Dr. Tolba, if the major economic powers and oil-rich nations follow suit, we will be in business. He sees the level of their funding, not in millions, but in billions of dollars. The call was recently echoed by Rajiv Gandhi, who suggested the creation of an environment protection fund.

This is the type of commitment and responsibility nations must display to meet the challenge of the global climate crisis. So far, the superpowers have not pronounced on a possible climate fund. Although, as I understand it, the European Community is examining the possibility of implementing a half a cent levy per kilowatt-hour of fossil fuel used, which could generate some $55 billion a year. Funds would be used for debt reduction and reforestation. Perhaps the superpowers might better seek to address problems related to climate change by building strategies of multinational and global cooperation.

The superpowers have the most advanced technologies, both in the terms of energy-efficiency and environment-benign innovations—such as alternatives to chlorofluorocarbons. What strategy for transfer of technologies have the superpowers considered? A willingness to share technologies on the part of the great industrialized nations would go a long way to enhance the confidence-building process upon which global consensus and cooperation must rest.

The United Nations is doing its part. The challenge now is "what the world's peoples can or want to do." The major responsibility of Earth's inhabitants is to create the demand for effective measures to understand and control and insist upon the climate change, and initiation of environmentally sound policies.

UNEP warmly commends *Greenhouse Glasnost* as a major step in the right direction, a very creative and timely initiative for positive change and multilateral cooperation.

THOMAS E. LOVEJOY

Will Unexpectedly the Top Blow Off?

Thomas E. Lovejoy is a tropical and conservation biologist who has worked in the Amazon of Brazil since 1965. He conceived the Minimum Critical Size of Ecosystems Project, for joint research by the World Wildlife Fund–U.S. and Brazil's National Institute for Amazon Research. For his scientific work and conservation initiatives, Dr. Lovejoy was decorated by the Brazilian government with the prestigious Order of Rio Branco.

In the field of international conservation, he is the originator of the innovative concept of "debt-for-nature" swaps. The first four such swaps were initiated in Bolivia, Costa Rica, Ecuador, and the Philippines, and others continue to be developed. Dr. Lovejoy is the creator of the PBS series *Nature*, chairman of the United States Man and Biosphere Program, president of the Society for Conservation Biology, and an executive member of the Scientific Committee on Problems of the Environment (SCOPE). He serves as a member of the President's Council of Advisors in Science and Technology and has written or edited many articles and books, including *Key Environments: Amazonia* (with G. T. Prance). Dr. Lovejoy is currently the assistant secretary for external affairs at the Smithsonian Institution.

I n Archibald MacLeish's poem "The End of the World," a circus crowd
is suddenly surprised when the big top blows off. But I would read into
this metaphor that the unexpected is a matter of perception, and that
attention to, as opposed to oblivious nonchalance about, incremental change
might have removed the element of surprise.

I have had a conviction for some time that the absence of widespread
understanding of fundamental concepts about how the world works is at the
heart of the inertia of the general public, and hence of government, to
address basic issues. Most particularly, a lack of appreciation for what
exponential increase really means leads society to be disastrously sluggish
in acting on critical issues. The financial community *must* understand exponential increase, for that, after all, is what Shylock's evil—interest
rates—and discount rates are all about; yet even so, the world of commerce
seems largely unable to extend it to the larger world within which it
operates. It is almost as if there were a desperate conservative desire for
the status quo built into our genes in futile defiance of the realities of
natural selection. Perhaps we evolved in a time of change slower than that
we are afflicting upon ourselves. Or perhaps this reflects the difficulty of
distinguishing between the initial phase of an exponential curve and random fluctuation.

In 1988, I found to my personal horror that I had not been immune to
naiveté about exponential functions in my active engagement on international environmental problems over the past fifteen years. It is more precise
to say that while I had been aware that the interlinked problems of loss of
biological diversity, tropical deforestation, forest dieback in the Northern
Hemisphere, and climate change are growing exponentially, it was only two
years ago that I truly internalized how rapid their accelerating threat really
is. If the change they represent is unacceptable (or at least mostly unacceptable), the rate of the change is more unacceptable, and its second derivative,
its acceleration, is even more unacceptable.

I am utterly convinced that most of the great environmental struggles will
be either won or lost in the 1990s, and that by the next century it will be
too late. That is not to say that all the undesirable change will have taken
place. Rather, by then the momentum of the problems, coupled with the
inertia of society, will render the problems insuperable. Consider for exam-

ple Alberto Setzer's report from the Brazilian Institute of Space Research. He estimated that in 1987, a total of 20 million hectares (almost 50 million acres) of Brazilian Amazonia burned, of which 8 million hectares (or roughly 20 million acres) were primary forest. This is not an activity one can turn off at the flick of a switch.

Perhaps my greatest sense of desperation comes from the specter of climate change. The greenhouse effect is upon us and the preview provided by the 1988 drought, itself perhaps only a normal fluctuation, heightens consciousness. Yet what does the response seem to be? A few still deny it and would appear to prefer to conduct the gigantic global experiment before they will believe—as if oblivious that it is unscientific to conduct an experiment without a control. Yet while the greenhouse effect is widely accepted now as a growing reality, some still refuse to accept it as a crisis, and others indulge in efforts I categorize as "dike mentality"—namely, to devise ways to treat symptoms, like developing new crop strains, for example—rather than to address the causes. I do not mean to belittle such efforts, for they have their place: but only as a part, and rather a minor part of what should be done. And there are still some who suggest, like the physicist who wrote a letter to the editor of *BioScience,* that the fertilizing effect from increased carbon dioxide levels in the atmosphere made the greenhouse effect rather desirable.

Indeed, there seems to be general public ignorance of how the greenhouse effect is linked to the biology of the planet. People do not understand:

- that the increase of carbon dioxide (CO_2) in the atmosphere comes in part from the burning of old biomass (e.g., fossil fuels) and in part from the destruction of modern biomass (e.g., forests in temperate and tropical regions),
- that the increase of CO_2 has great significance for the biota,
- that biology can be used to avoid an important portion of the greenhouse effect.

Carbon, as CO_2, is accumulating in the atmosphere at an annual rate of 3 to 3.5 billion tons. This accumulation is not a matter of dispute; it is measurable. That the two major sources of the release of carbon are fossil fuels and modern biomass is not a matter of scientific contention. The debate is over the relative portion of each, as well as the extent of importance of the oceans as a sink for carbon. While more precise understanding is needed, we must not allow ourselves to be distracted by what in a sense could be a "precious" academic debate. What is clear is a "bottom line" of CO_2 increasing in the atmosphere.

The consequences on climate are more complex and less well understood. It seems to be generally agreed that temperature change will be uneven with respect to latitude; there will be little difference in the tropics and great changes at higher latitudes, with the horrifying possibility of increased release of methane in tundra regions—in short, a runaway greenhouse effect. A 2°C global increase can convert to changes of 5° to 10°C in higher latitudes. The prospect of additional geographical unevenness superimposed within the temperate zone could produce even more extreme changes locally. It should not be concluded that the tropics will be immune to change. The shifts there are more likely to involve rainfall patterns.

The implications of the greenhouse effect may or may not be deeply disturbing to those who study cycling processes of ecosystems and perforce study them as if species-blind. From the perspective of biological diversity, however, climatic change is deeply worrying, because much of temperate-zone diversity and some boreal diversity is increasingly restricted to a small number of national parks and biological reserves. If the climatic conditions to which species are suited move away from their reserve areas, will they have anywhere to follow in our man-dominated landscapes? In those situations where such species do have somewhere to go, will they be able to move rapidly enough? Paleoecological work such as that of Margaret Davis can be of some instruction here, but the fundamental question about all kinds of biological responses to the greenhouse effect is: Can it match the projected rate of climatic change? This was examined in a workshop sponsored by World Wildlife Fund, the Smithsonian Institution, and the National Science Foundation in October 1988. The workshop could not yield a tremendous amount in the way of clear answers but it definitely added loss of biological diversity to the many environmental concerns about the greenhouse effect.

Even without worries about biological diversity, other environmental consequences are cause for major alarm *and* action. Rising sea levels will threaten coastal cities, fertile agricultural deltas, and at least temporarily, coastal wetlands. There will be a major reordering of agriculture in temperate zones as new climatic and soil combinations occur. In the tropics, possible rainfall shifts will render the subsistence farmers vulnerable, and agricultural extension systems that could supply new crop strains are often rudimentary at best. There will be some benefits, of course, such as an extended wheat-growing season in Siberia. In sum, however, there will be no winners in this global game of ecological chairs, for it will be fundamentally disruptive and destabilizing; we can anticipate hordes of environmental refugees dwarfing the numbers of Dust Bowl migrants or boat people.

The only acceptable response is to bring the global carbon cycle to a state akin to normal. The first step is to eliminate the annual increase of atmospheric carbon dioxide. This can be done by a combination of reducing CO_2 emissions (through increased energy efficiency, use of alternate energies, and control of forest dieback in northern regions and of tropical deforestation in the southern lands) and increasing the amount of carbon fixed through reforestation. An area of 2 million square kilometers of newly established forest will take up approximately 1 billion tons of carbon annually until the new forest reaches maturity. What this buys is time—time to develop a better management of energy use and reduction of dependence on carbon-based fuels. This is not really a large area—only 1,414 kilometers on a side—and obviously does not need to be all in one place. There are plenty of degraded lands around the world that could be devoted to the purpose without any conflict with agricultural needs. It should even be possible to make at least a small annual reduction in the atmospheric level of CO_2 accumulated since the Industrial Revolution, although *removal* at a rate of 1 billion tons per year would attain eighteenth-century levels only in 152 years. This may seem inconsequential, especially in light of likely increases in energy demand, but other sinks for carbon may be possible and technological advances such as miniaturization and superconductivity can help with the equation.

The only solution I see is a global protocol, probably a convention, to manage carbon and forests on a global basis in order to stabilize the composition of the atmosphere. The recent success along these lines with ozone production demonstrates that this can be done in principle. These would not be easy negotiations, to say the least, for energy and carbon are central to modern society and aspirations of less developed nations for development need special concern. Yet already at least one country, Switzerland, by law maintains minimum forest cover, and there probably are interesting lessons to be derived and applied to the global situation. Further, one far-seeing American power company has undertaken a reforestation project that will take up an amount of carbon equivalent to that the company anticipates releasing from a new coal-fired plant. Parallel efforts will need to be conducted, of course, for the other greenhouses gases and other global cycles, but carbon is the most important.

The role of technology must be considered. It is true that technology in part got us to this state, driven by population increase and the world of commerce. It would be a mistake, however, to recoil in catatonic technophobia, because new technologies better integrated with the natural world are a critical part of the solution. I do not mean a technological fix akin to Venetian blinds against the sun, as has actually been proposed, but rather technologies that will help reduce the release of carbon dioxide into the atmosphere. Any other means of reducing the greenhouse effect is as dangerous as the fiddling that created the problem. It must always be borne in mind that the biosphere and geosphere are extraordinarily complex, with somewhere between 10 and 30 million interdependent living parts called species, and numerous hard to detect threshold effects. Solutions respectful of this amazing complexity are more likely to be benign than a simple engineering attempt to lower the thermostat.

The problem of tropical deforestation and the staggering associated loss of biological diversity must not be overlooked in a preoccupation about climate. Every fresh view seems to increase the estimate of total number of species on the planet, and the percentage of total diversity in the tropical forests, which are at grave risk. With a seriously imprecise measure of global diversity, but with a sense that from 50 to perhaps 90 percent of all species are in the 7 percent of the Earth's dry land area represented by tropical forests—which are vanishing at well in excess of 30 hectares (74 acres) per minute—clearly the problem is very big and the fuse very short. On top of the massive and accelerating extinctions engendered by sheer destruction, there is the additional concern about regional climate change (as in the work of Eneas Salati, former director of Brazil's National Institute of Amazon Research and the Center for Nuclear Energy and Agriculture, on the Amazon hydrological cycle) and its impact on tropical biological diversity. The greenhouse effect may add an additional burden, for although temperature will be little affected in the tropical zone, rainfall may be; seasonality and totality of precipitation are fundamental environmental factors in the tropics.

All this is happening at a time when the importance of biological diversity to society grows daily. Yet this transpires largely in unpredictable ways, which force us to argue for biological diversity either in generalities or with vivid examples. That is to say, serendipity plays a large role in science, and particularly biological science. This, of course, is part of the fun of it, but to admit this publicly can seem terribly *unscientific:* however rigorous the application of the scientific method, there will be many surprises. Charles Goodyear's discovery of vulcanization, Leo Hendrik Baekeland's of plastics, and the discovery of Teflon's many modern applications were all accidental. The vast majority of plant and animal species and the tremendous complexity of living systems are not described by science, and so the role of serendipity in the ecological sciences must be overwhelmingly even greater.

Who among the eighteenth-century devotees of the glories of the French vineyards, such as Thomas Jefferson, would have guessed that by the end of the next century they would all depend on wild American rootstock? Who twenty years ago would have believed there could be biological communities running on the primal energy of the earth, that the African clawed frog would yield important healing compounds, or that a South American pit viper that executes with a venom that permanently lowers blood pressure would reveal an entirely new system of blood-pressure regulation in humans (from which a pharmaceutical company makes hundreds of millions of dollars a year)?

The point to be made here is that biologists have a vested interest in maintaining the fullest possible biological diversity, as indeed does society in general. If knowledge is power, then protecting the basis for building knowledge about life on earth is the basis for power—power to benefit from the biological dowry of our planet and to manage it and ourselves wisely. Biologists should not be ashamed to admit the major role of serendipity in their science even if they are derided by some physical scientists who are nonetheless subject to it themselves. Rather, it becomes a strong argument to society in favor of conserving biological diversity, and recalls a remark of Louis Pasteur: "In the fields of observation chance favors only the mind that is prepared." Never has conservation needed science more, but in a newly energized and thoughtful sense. More science-as-usual is not a solution in itself; additional lines of inquiry and new approaches are needed.

Systematic biology (which describes the variety of life on the Earth) is a declining science at the moment, but it has an enormous role to play in the conservation of biological diversity. If 95 percent of all species are yet to be described, economic development and conservation—and for that matter, science—are truly stumbling around in the dark. Yet entomologists, who have the greatest amount to lose in terms of diversity, tropical or otherwise, seem little stirred as a group by problems related to biological diversity. At a recent international congress of entomologists held in Vancouver, there was only one symposium on conservation, and it had to compete with twenty other concurrent sessions. How can complete understanding about Haldane's musing about God's inordinate fondness for beetles be achieved without the full coleopteran panoply? Shouldn't we all demonstrate an inordinate fondness for beetles, and indeed for the Creation?

It is ironic that those scientists specializing in one group of plants or animals are often the most fervent defenders of endangered species, but their product is ill-suited to the conservation task, which is immediate and immensely practical. Great monographs and elaborate phylogenies inherently take too long to provide the biogeographical map of life on Earth needed for conservation and development to proceed effectively and wisely.

More money for systematics-as-usual, while warranted on other grounds, is not the solution for problems related to biological diversity. Nor is the equally valid goal, articulated by Ed Wilson among others, of a complete inventory of biological diversity. What is needed is a global plan to sketch in the biogeography of the planet rapidly. With diversity conserved, the monographs and phylogenies can be produced over centuries to come, as indeed can the "Compleat Inventory." Such a global crisis–oriented project will not come about without a general stirring of organism biologists. What we are talking about is not rejecting science but rather, in the mode of the report "Research Priorities in Tropical Biology," making research choices, getting on with the work, and disseminating results by the most modern and rapid means possible.

A recent National Science Foundation-sponsored meeting considered how the newly emerging science of conservation biology, now represented by an adherent society, could contribute solutions. The report is a recipe for an epidemiology of the planetary environment. Many of the suggested lines of research are no surprise, such as the need for a variety of mega-experiments like the Minimum Critical Size of Ecosystems Project in Amazonian Brazil. Of particular interest is the greater awareness of the importance of the social sciences as conservation biology gropes to understand the causes and consequences of, and ultimately the proper responses to, our effect on the environment. The Man and the Biosphere Program takes on an enlarged relevance.[1]

In this time of the crisis of life on Earth it is vitally important that we avoid divisiveness and work together to advance relevant science and its funding. It would be very destructive if particular branches of science are perceived to be out to feather their research nest instead of being preoccupied with the larger problem. It is important also that scientists as a group are not viewed as singing only the self-interested refrain of the need for more research. We must internalize the reciprocity of science and conservation to the point where we are seen universally as active in science *and* conservation. The only way to deal with the monumental problem is by approaching it on the same scale and in a holistic fashion. I am not suggesting that scientists dissipate their energies and effectiveness, that everyone try to do everything; but by and large, scientists are perceived as a mysterious if not odd lot, addicted to the pleasures of intellectual pursuit. To avoid misinterpretation as they ask for more funds and big changes in

1. Man and the Biosphere is an interdisciplinary coalition of scientists supported by the U.S. Department of State and other government agencies. It sponsors policy-relevant interdisciplinary research projects and provides legislative policy guidance to "foster harmonious relationships between humans and the biosphere."

the way society works, scientists need to articulate the entire context of the problems caused by a global warming.

When it comes to the international dimension, we need to be acutely sensitive to the social needs of tropical nations and to building local scientific capacity. There is no room here for any form of arrogance. Nothing short of a full global partnership will prevent these terrible problems. Indeed, how and how well the richest nation on Earth handles its own diversity problem sets a critical example for less wealthy nations.

The crisis of life grows ever greater and ever more insistent, like an environmental *Bolero*. This is a time for scientific and environmental statesmanship of the highest order. This is in many ways the moment in history for biologists. Daunting as the task may be, we should set our sights at the highest possible point.

We must explain that the solutions, while seeming outrageous, are in fact reasonable when the consequences are contemplated. Reordering of the international debt, for example, is a tantalizing prospect when one considers the economic reordering inherent in some of the solutions science does and will present. Biologists must not leave the problems of global climate change merely to the technologists or physicists to solve. The solutions are technological and physical only in part, and some physical scientists do not understand biological problems sufficiently to know what they really are.

We, humanity, have finally done it: disturbed our environment on a global scale. Scientists and biologists have a responsibility—an imperative—to speak out, or all of society's activities will be played out on a global Okefenokee of shifting environmental ground. Before I had truly realized the scope of the problem, I used to think the tragedy was that so much would be lost and nobody would notice a century from now. I now believe the impact on society could be so great that it will be remembered if we do not speak out and act. The alternative is to go our merry intellectual ways, pleasuring ourselves with fine new theories and hypotheses, self-indulgent and distracted, rather like the circus crowd in MacLeish's poem, when

Quite unexpectedly the top blew off:

And there, there overhead, there, there, hung over
Those thousands of white faces, those dazed eyes,
There in the starless dark the poise, the hover,
There with vast wings across the cancelled skies,
There in the sudden blackness the black pall
Of nothing, nothing, nothing—nothing at all.

GEORGE M. WOODWELL

Forests and the Atmosphere in a Warming World

George Masters Woodwell is founding director and president of the Woods Hole Research Center, a global environmental research institute. Dr. Woodwell's work as a biologist focuses on the structure and function of natural communities and their role as segments of the biosphere. He has worked in the forests of North America, in estuaries, and has made a special study of the ecological effects of ionizing radiation, pesticides, and other toxins. Currently he is involved in an analysis of the function of biota in the world carbon budget. Dr. Woodwell has published and has edited books on the effects of nuclear war, the global carbon cycle, and satellite imagery applied to measuring the area of forests worldwide. He is a founding trustee of the World Resources Institute, a former president of the Ecological Society of America, and former chairman of the World Wildlife Fund. He is a fellow of the American Academy of Arts and Sciences, and a member of the National Academy of Sciences.

FORESTS AND CLIMATIC CHANGE

Deforestation is both cause and effect of the accumulation of heat-trapping gases in the atmosphere. Deforestation results in the accelerated decay or burning of organic matter of plants and soils, with the release of carbon dioxide and methane into the atmosphere. The warming induced by the accumulation of such heat-trapping gases causes the further destruction of forests and their soils. How does this process work, and how serious are the consequences?

The Warming of the Earth

If climatologists are correct and well-established physical principles hold, the Earth will warm over the next years by a global average of 1.5° to 5.5°C because of the accumulation of heat-trapping gases in the atmosphere (S. Schneider, 1990; Smith and Tirpak, 1989; WMO/UNEP, 1988). Climatologists do not say the only factor affecting the temperature of the Earth is the composition of the atmosphere. Nor do they say the dominant effect on temperature is always caused by heat-trapping gases. They do point out that adding these gases to the atmosphere will warm the Earth above whatever its temperature would have been without them. The rate of accumulation is high enough to suggest there will be a sudden warming over the next decades. The warming may be associated with other, less predictable but sudden changes in regional climates in response to changes in the circulation of the oceans or other factors (Street-Perrot and Perrot, 1990).

If ecologists are correct and well-established ecological principles hold, a rapid global warming will bring the further release of carbon, as carbon dioxide and methane, from tundra and forests; this release will speed the warming. It is difficult to estimate the magnitude, but it is possible that several billion tons of carbon annually will be added to current releases, and such a release may be expected to push the warming toward or beyond the upper limit defined by climatologists. There is evidence that such releases have occurred in the past and are occurring now (Houghton and

Woodwell, 1989; Woodwell, 1983, 1989), and that other factors may accelerate the warming trend (Kellogg, 1983; Lashoff, 1989; Houghton et al., 1985).

The accumulation of carbon dioxide and methane in the atmosphere has been caused in part by the destruction of forests. Deforestation was the dominant source of carbon dioxide accumulation in the atmosphere until the middle 1960s, when a global surge in the use of fossil fuels brought emissions from them above the releases from deforestation for the first time (Woodwell et al., 1983). In the late 1980s, deforestation was thought to be occurring at a higher rate than ever before, but the surge in use of fossil fuels continued and was thought to exceed the release from deforestation by a factor of approximately 2. Emissions from fossil fuels in 1990 will exceed 5.6 billion tons of carbon; releases from deforestation will probably be in the range of 2 to 3 billion tons.

Forests have been destroyed continuously throughout history. The last of the world's primary forests are being destroyed now, usually as a matter of governmental policy. It is the policy of Brazil to encourage agriculture and other enterprises in the Brazilian Amazon, all of which depend on destruction of the forest. Blatant corruption has led to the rapid deforestation of the Philippines by commercial interests. The developed nations are not better. Both Canada (Shea, 1990) and the United States are actively engaged in the destruction of their residual publicly owned old-growth forests. The subsidized destruction of the Tongass National Forest in Alaska has become a national scandal. Responsibility here lies heavily with Alaska's Senator Murkowski, but he is not alone. In this instance, instead of offering leadership to the world, the United States offers the very example of destructive exploitation objectionable in Brazil and elsewhere, where the destruction of forests is leading rapidly to impoverishment of the biota, the land, and the people.

The prospects for forests are grim under the best of present conditions. Tropical forests are being destroyed at rates that leave even foresters, long experienced in turning trees to sawdust, bewildered. Forests globally are vulnerable to demands for lumber, pulp, land for farming and grazing, and to speculation in real estate in circumstances where ownership is established by clearing the land. These objectives compete with the acidification of rain, the toxification of air and water, and with the effects of mining, to destroy the remaining primary forests and often the potential of the land for supporting life.

As if to complete this destruction once and for all, eliminating any prospect of serious human reliance on forests as opposed to complete enslavement by the industrial system, we now have the prospect that climatic changes globally will destroy forests over large areas. The anticipated changes in climate are, to be sure, uncertain—and may proceed at rates much lower than the maxima suggested by climatologists. On the other hand, there are reasons to believe those maxima were deduced without consideration of several factors that will speed the warming as its effects begin to be felt (Kellogg, 1983; Lashoff, 1989; Woodwell, 1983). The acceleration could quickly push the warming into the upper range of rates anticipated by climatologists' models based on the immediate past—and might push it into a realm well beyond current predictions. The cause of the acceleration is the rapid release of carbon dioxide and methane from forests and tundra in particular. A recent appraisal of the availability of methane in clathrates in shallow seas and in the Arctic suggests a very large potential for a further release as the warming progresses (MacDonald, 1988). These changes apparently have dominated temperature variations during the glacial age (Lorius et al., 1988, summarized in Houghton and Woodwell, 1989), although the causes of the shifts between warming and cooling trends have not been defined. The prospects of a global warming to approach or exceed in rate the highest estimates climatologists have yet produced seem sufficiently real to require serious consideration. How important would forests be in such changes, and how would they fare on a rapidly warming earth?

Forests and Climate: Carbon Dioxide and Methane

The role of forests in affecting the atmosphere's composition has been grossly underestimated by climatologists, largely as a result of too much focus on oceans. The most conspicuous short-term effect is the control forests exert over the carbon dioxide content of the atmosphere, as recorded in data reporting seasonal trends at the Mauna Loa Observatory in Hawaii and elsewhere. At Mauna Loa during the summer months, photosynthesis of forests in the Northern Hemisphere becomes the dominant influence and reduces the carbon dioxide content of the atmosphere by about 5 parts per million. During winter, respiration dominates and restores carbon dioxide removed during summer. The process has been summarized by Woodwell (1983), who showed that forests control these shifts. Globally, there is an annual flux of carbon through plants, animals, and decay organisms on land of about 100 billion tons, approximately an eighth of the atmospheric burden. Any modification of that flux has the potential for affecting the composition of the atmosphere appreciably. Most of that flux is through forests simply because forests are so large in area and carry on such a large fraction of the global metabolism on land. There is also a large stock of carbon in the plants and soils of forests that can be mobilized as carbon dioxide released into the atmosphere (Woodwell and Pecan, 1972). Changes in the stocks of carbon in forests—including their soils and the atmosphere—and in the fluxes between them can occur rapidly in response to climatic changes, or even changes in weather.

The most important biotic changes will involve immediate shifts in the ratio of gross production (total photosynthesis) to total respiration locally, regionally, and globally. A global warming will affect the balance between the approximately 100 billion tons of carbon entering green plants annually through photosynthesis on land, and the 100 billion tons released through respiration. The annual net accumulation of carbon in the atmosphere has been about 3 billion tons over the past fifteen years, and a 1-percent change in either process, photosynthesis or respiration, will have important implications for the Earth's atmospheric composition in a short time. The question of how and under what conditions this ratio is affected reaches to the core of the evolutionary process and raises issues that often leave mere mortals floundering.

Nonetheless, answers seem to be forthcoming. The most powerful clues come from analyses of glacial ice laid down over 160,000 years of the Earth's recent past (Barnola et al., 1987; Lorius et al., 1988). Data show the carbon dioxide content of the atmosphere increased when temperature was rising and declined during cooling periods. The correspondence was not perfect: there appears to have been a lag during cooling periods in the rate of decline of carbon dioxide content. But the pattern is consistent with the hypothesis of a positive feedback through biotic systems: A warming makes the warming worse, and a cooling withdraws carbon dioxide from the atmosphere and makes the cooling worse.

The origin of this feedback system seems to lie in the effect of changes in temperature on the storage and release of carbon in terrestrial ecosystems, especially forests. There is strong evidence that the dominant effect is the influence of temperature on respiration. A change of 1°C commonly alters the rate of respiration by 10 to 30 percent, sometimes more. An equivalent change in the rate of gross photosynthesis is difficult to effect, and although there may be a change in response to the enrichment of the atmosphere with carbon dioxide, increased precipitation, or a longer growing season, the stimulation of respiration is greater than the total of all other effects. The net result of a warming is the release of more carbon into the atmosphere. The process seems to have been active throughout the last 160,000 years while glaciers were advancing and retreating in reaction to temperature changes.

The same process seems to be active now. During the past twenty-four months the rate of accumulation of carbon dioxide in the atmosphere, measured as carbon, has increased from about 3 billion tons annually to about 5 billion, apparently as a result of the increase in the Earth's temperature in the 1980s (Houghton and Woodwell, 1989; Woodwell, 1989).

While the circulation of the oceans may be critical in determining climates regionally, and the absorption of carbon dioxide by the oceans remains a significant factor in determining its year-to-year concentration in the atmosphere (Houghton et al., 1985), the pool of carbon in forests is both large and volatile. Shifts in that pool occur rapidly enough to give forests the dominant role for months and even years in determining the atmospheric burden of carbon dioxide.

Methane too is a product of the decay of organic matter in soils, especially in soils that are periodically waterlogged, and where a significant fraction of the decay occurs anaerobically. Such circumstances occur in forest soils at all latitudes, especially in the higher ones, and in the tundra. It is probable that a significant fraction of the methane now accumulating in the atmosphere comes from soils as a result of the stimulation of the respiration of decay (Khalil and Rasmussen, 1989) by the warming that has already occurred—0.5° to 0.7°C over the past century, according to Hansen and Lebedeff (1988). Again, any step that reduces deforestation or further warming of the Earth will contribute to the control of methane and reduce its concentration in the atmosphere.

I have emphasized interaction of forests with the warming of the Earth now under way. The interactions are large and important and can be defined quantitatively, but they offer only one of the bases for analyses of the pervasive role forests play in maintaining the human habitat. The erosion of the normal function of forests and other natural communities is incremental. When complete, not only are previously existing biotic resources destroyed, but through the loss of nutrient elements from the land and erosion of topsoil, the potential of the land to support life is reduced as well. The impoverishment spreads to the populace, which finds a reversal of the process, however desirable, virtually impossible for a host of biotic, political, and economic reasons. Examples abound: coastal Brazil, Madagascar, Haiti, Mexico, India, and a score of other countries.

India is a compelling example. Governmental officers currently assert that one-third of the land area of this populous country is impoverished to the point that it supports neither forests nor agriculture, including grazing areas. Some of that land is rock, until recently covered by a thin soil, now eroded by a combination of overgrazing, wind, and monsoonal downpours. Some is salinized but potentially recoverable under appropriate, albeit novel, management. Some is simply overgrazed to the extent that vegetation has been eliminated or shifted to hardy species that are not nutritious (Woodwell, 1990). Under these conditions, the process feeds on itself and the soil often becomes saline at the surface, and sterile. Irrigation almost always leads to salinization and impoverishment of soil.

The process of impoverishment varies little around the world. In the tropics, where certain soils are particularly vulnerable to loss of nutrients, it may proceed more rapidly—and recovery may be more difficult, even impossible. Brazil is currently encouraging the migration and settlement in the Amazon basin of thousands of its citizens, displaced from industrialized agriculture in the south. The settlers clear forested land, farm it successfully for two or three years until crops fail on soils that have low capacity for retaining nutrients. Speculators buy the land for the grazing of cattle at very low density, and the settlers move on to clear a new tract. The effect is to destroy one of the oldest and richest self-renewing biotic systems on Earth and replace it with one of the most impoverished, useful only for the grazing of cattle at one steer per 1 to 10 hectares (2.47 to 25 acres). It is ironic that the harvesting of rubber alone from the intact forest yields a greater income than the impoverished grazing land. And the potential of the intact forest for further renewable use has only been scratched. In addition, destruction of the forest releases nutrients into streams, speeds runoff from the land, hastens erosion; and the streams enter the cycle of eutrophication and impoverishment as well.

The process is not limited to the tropics. Similar transitions are common in other latitudes, but they may be less conspicuous and less severe where soils are younger, and have a greater capacity for retaining or supplying nutrients, and where seasonal swings of weather limit the distribution and effect of plant pathogens. The impoverishment of land has economic implications that are largely unmeasured, however important and obvious. Any reduction in agricultural land in a country as populous as India presents an obvious burden on the nation as a whole, raises prices domestically, and reduces the possibility of exports important in producing foreign exchange. But more than that, wood for fuel is scarce and becomes even more so as larger areas become impoverished. A shift to oil, purchased on the international market, is one solution suggested for the shortage of fuel wood.

Solutions that rely on exogenous products introduce a need for cash into an economy in which cash is short. In this instance, because oil is not produced in India, using it to replace firewood only adds to foreign-exchange problems. Biotic impoverishment is clearly linked to economic impoverishment, personally, nationally, and globally.

REFORESTATION

Can reforestation be used to remove carbon from the atmosphere, at least temporarily?

Simple answers abound. The simplest is the common reaction: Plant trees. While the suggestion is constructive, the goal is to remove carbon from the atmosphere and store it for decades to a century or more. Such a goal requires the reestablishment of forests. Merely planting trees does not build forests with the capacity for storing carbon in soils as well as plants, for absorbing and recirculating nutrient elements, and for regenerating after disturbance. Establishment of plantations is a step in the right direction, but plantation forestry is primitive in most applications; the forests are vulnerable to fire and disease; and the expense of planting is great enough that the forests must be harvested systematically to warrant the investment. Current reasons for establishing plantations are antithetical to the objective: accumulating carbon stocks in forests and soils, and preserving those stocks over time. Plantations are commonly begun with early harvest in mind, not with the aim of reestablishing a fundamental biotic unit to serve multiple functions in stabilizing the landscape, in preserving biotic resources, and in helping to control the Earth's atmospheric composition. Building forests designed to persist for a century or more requires greater knowledge and effort than is usually available for establishing tree plantations.

The issue of reforestation is more complicated still. An area of 1 to 2 million square kilometers of forest is required to store 1 billion tons of carbon annually (Marland, 1988; Woodwell, 1989). Impoverished land is no better for forests than for agriculture. The richer the land, the more rapid the growth of forest plants and the greater the storage of carbon. Since it is unlikely that large areas of productive land can be made available for reforestation—most productive land in countries such as India is used for intensive agriculture—clearly the first objective in management of forests must be an end to deforestation. Efforts at reforestation seem destined to play a secondary role, however attractive they may appear initially.

Nonetheless, throughout much of the world abandoned land is reforested in time through normal, natural successional processes; these have been at the core of much research and analysis in ecology (Clements, 1928; Oosting, 1958). Succession, however, requires that several conditions be met: sources of seeds must be available, and soil, mineral nutrients, and climate must be suitable. In large areas of the tropics deforestation leads quickly to degradation of soils and loss of nutrients, and it may lead to such profound changes in the site that it will no longer support forests. Once destroyed over large areas, forests do not necessarily recover naturally. Massive programs of reforestation will be expensive, difficult to execute, and probably slow to become effective. Forests are in competition with agriculture in a world where the human population is expanding at 90 million people annually, and with open land that seems to offer more immediate financial potential in the cultivation of other crops. Agricultural crops do not have the potential for balancing continued releases of carbon from the global destruction of remaining primary forests, and cannot— under any conditions imaginable at the moment—have the potential for balancing continued use of fossil fuels as the primary source of energy for an expanding industrialized world.

AN ALTERNATIVE

The time when forests in their entirety can be considered local, regional, or even national resources to be managed for personal or community interests has passed. They are an essential segment of the biosphere: essential to the stabilization of climate globally; to the management of water, land, and air regionally and locally; and to the preservation of the terrestrial biota. Forests are a part of the global commons, essential to the welfare of all, now and for the future.

We have little experience in managing supranational resources and less in recognizing and facilitating the transfer of private or national resources in part or in whole to global status. Yet such a transfer is what is called for now in the common interest. Can it be effected?

This challenge will require leadership from science and government. Basic questions lie unanswered and unaddressed: How much forest is required to stabilize the composition of the atmosphere? To provide for the preservation of the Earth's biota? To provide for the stabilization of nutrient fluxes and water flows? To provide for the world's needs for fiber and lumber and the diversity of products, actual and potential, from the world's forests? What is the best way to manage the Earth's gene pools to prevent their being systematically impoverished?

And if answers were available, as they *will* be, how might we proceed toward enacting systems of control to assure that we use the flow of resources sustainably and do not rashly invade the capital?

If there is a need for a global convention to address the challenges created by the warming of the Earth, there is a parallel need for a global convention to address the management of forests and to assure that present trends do not result in rapid and irreversible impoverishment accompanied by progressive destabilization of climate worldwide.

REFERENCES

Barnola, J. M., D. Raynaud, Y. S. Korotkevich, and C. Lorius, "Vostok Ice Core Provides 160,000-Year Record of Atmospheric CO_2," *Nature* 329, no. 6138 (1987), pp. 408–414.

Clements, F. R., *Plant Succession and Indicators.* Washington, DC: Carnegie Institution, 1928.

Houghton, R. A., W. H. Schlesinger, S. Brown, and J. F. Richards, "Carbon Dioxide Exchange Between the Atmosphere and Terrestrial Ecosystems," in J. R. Trabalka (ed.), *Atmospheric Carbon Dioxide and the Global Carbon Cycle.* Springfield, Va.: National Technical Information Service, 1985.

Houghton, R. A., and G. M. Woodwell, "Global Climatic Change," *Scientific American* (April 1989), pp. 36–44.

Kellogg, W. W., "Feedback Mechanisms in the Climate Systems Affecting Future Levels of Carbon Dioxide," *Journal of Geophysical Research* 88: (1983), pp. 1263–1269.

Khalil, M. A. K., and R. A. Rasmussen, "Climate-Induced Feedbacks for the Global Cycles of Methane and Nitrous Oxide," *Tellus* 41B (1989), pp. 554–559.

Lashoff, D., "The Dynamic Greenhouse: Feedback Processes That May Influence Future Concentrations of Atmospheric Trace Gases and Climatic Change," *Climatic Change* 14 (1989), pp. 213–242.

Lorius, C., N. I. Barkov, J. Jouzel, Y. S. Korotkevich, V. M. Kotlyakov, and D. Raynaud, "Antarctic Ice Core: CO_2 and Climatic Change over the Last Climatic Cycle," *Eos* (1988), pp. 681–684.

MacDonald, G. J., "Role of Methane Clathrates in Past and Future Climates." Paper presented at the Second North American Conference on Preparing for Climatic Change: A Cooperative Approach. Washington, DC, 1988.

Marland, G., *The Prospect of Solving the CO_2 Problem Through Global Reforestation.* Washington, DC: U.S. Department of Energy, 1988. (NBB-0082 ORNL.)

Oosting, H. J., *The Study of Plant Communities.* San Francisco: Freeman, 1958.

Schneider, S. H., "The Changing Climate: A Risky Planetwide Experiment," *Greenhouse Glasnost* (T. J. Minger, editor). New York: The Ecco Press, 1990.

Shea, C. P., "The Great Northern Forest Sell-off," *World-Watch* 3 (1990), pp. 8–9.

Smith, J. B., and D. Tirpak (eds.), *The Potential Effect of Global Climate Change on the United States.* Report to Congress. Washington, DC: U.S. Environmental Protection Agency, 1989.

Street-Perrot, F. A., and R. A. Perrot, "Abrupt Climate Fluctuations in the Tropics: The Influence of Atlantic Ocean Circulation," *Nature* 343 (1990), pp. 607–612.

WMO/UNEP, *Developing Policies for Responding to Climatic Change: A Summary of Discussions and Recommendations of Workshops Held in Villach and Bellagio, 1987.* 1988.

Woodwell, G. M., "Biotic Effects on the Concentration of Atmospheric Carbon Dioxide: A Review and Projection," in *Changing Climate.* Washington, DC: NAS Press, 1983.

————, "The Warming of the Industrialized Middle Latitudes 1985–2050: Causes and Consequences," *Climatic Change* 15 (1989), pp. 31–50.

———— (ed.), *The Earth in Transition: Patterns and Processes of Biotic Impoverishment.* New York: Cambridge, 1990 (in press).

Woodwell, G. M., J. E. Hobbie, R. A. Houghton, J. M. Melillo, B. Moore, B. J. Peterson, and G. R. Shaver, "Global Deforestation: Contribution to Atmospheric Carbon Dioxide," *Science* 222 (1983), pp. 1081–1086.

Woodwell, G. M., and Pecan (eds.), *Carbon and the Biosphere.* Washington, DC: U.S. Atomic Energy Commission, 1972.

U.S. SENATORS
JOHN HEINZ AND
TIMOTHY E. WIRTH,
AND ROBERT N. STAVINS

Innovative Policies for Sustainable Development in the 1990s: Economic Incentives for Environmental Protection

John Heinz was first elected to the U.S. Senate in 1976, after serving two terms in the U.S. House of Representatives. Senator Heinz is a ranking member of four Senate committees: Finance; Banking, Housing and Urban Affairs; Government Affairs; and the Special Committee on Aging. He has worked extensively on matters of retirement, federal health spending on Medicare and Medicaid, international trade, human development, and the environment. Senator Heinz cowrote the Pennsylvania Wilderness Act to create the state's only wilderness area. He has supported a variety of environmental legislation, including the law establishing

Superfund, the Leaking Underground Storage Tanks Trust Fund, and the Clean Water Act.

Timothy E. Wirth, U.S. senator from Colorado, specializes in environmental, arms control, and budget policy, and has authored major Senate legislation on global warming. He is a member of four Senate committees: Armed Services, Budget, Banking, and Energy and National Resources. He served six terms as a U.S. representative, from 1974 to 1986, and led efforts that dramatically changed U.S. telecommunications policy. A White House Fellow under Lyndon B. Johnson, he has served as a board member of Planned Parenthood and Denver Head Start. Senator Wirth devotes special attention to Colorado's conservation and environmental issues, and has been a leading spokesman for stronger federal clean-air laws. He is national chairman of the Alliance to Save Energy.

Robert N. Stavins is an assistant professor of public policy at the John F. Kennedy School of Government at Harvard University. Professor Stavins is an economist and is primarily interested in natural resources and environmental policy. His current research is focused on alternative cost-effective strategies for mitigating global climate change, the diffusion of pollution control technology, and land-use changes in the northeastern United States. Professor Stavins has served as a consultant to a number of environmental organizations, including the Environmental Protection Agency, the National Academy of Sciences, and the U.S. Department of the Interior. He is also a university fellow of Resources for the Future in Washington, D.C.

INTRODUCTION

During the past two decades, a host of environmental laws and regulations have been enacted in industrialized nations, and substantial gains have been made in environmental protection: in many spheres, the environment *is* cleaner now than it was before. But the United States and the world at large continue to face major environmental threats—both ongoing problems, such as local smog, groundwater pollution, and regional acid rain, and newly recognized problems, including the awesome possibility of global climate change. As we enter the 1990s, increased attention has been given by political leaders on both sides of the Atlantic to a promising set of new policies that recognize market forces, not only as part of the problem but also as a potential part of the solution.

In the United States, the nature and tone of political debate have evolved rapidly, culminating with the Bush administration's proposal in the early summer of 1989 for a major overhaul of the Clean Air Act to include a market-oriented approach to controlling acid rain.[1] Our report *Project 88* described a variety of innovative measures to enlist the forces of the marketplace and the ingenuity of entrepreneurs to help deter pollution and reduce waste and degradation of natural resources.[2] Practical employment of economic forces to achieve heightened environmental protection at lower cost to society was emphasized throughout.[3]

In the United Kingdom, the Thatcher government has embraced[4] a study

1. On June 12, 1989, President George Bush proposed a "tradeable permit system" for acid-rain control as part of the administration's amendments to the Clean Air Act. This proposal was sent to Congress on July 21, 1989, as Title V of the administration's bill.

2. Stavins, Robert N., ed., *Project 88: Harnessing Market Forces to Protect Our Environment—Initiatives for the New President,* a public policy study sponsored by Senator Timothy E. Wirth, Colorado, and Senator John Heinz, Pennsylvania (Washington, DC, December 1988).

3. Several studies have followed *Project 88;* see for example: John L. Moore, et al., *Using Incentives for Environmental Protection: An Overview,* Congressional Research Service Report to the Congress #89-360 ENR (Washington, DC, June 1989); and Robert C. Anderson, et al., *The Use of Economic Incentive Mechanisms in Environmental Management,* draft report (Washington, DC: American Petroleum Institute, September 1989).

4. See Nicholas Schoon, "Markets in 'Permits to Pollute' Proposed," *The Independent* (London), August 16, 1989; and Marion Shoard, "Clearing the Air with a Tax," *The Times* (London), September 15, 1989.

directed by David Pearce of University College, London, which recommends increased reliance on economic-incentive mechanisms for a host of environmental and natural resource problems.[5] Most recently, as massive political and economic changes have gripped the Soviet Union and Eastern Europe, a number of Eastern Bloc nations have expressed interest in market-oriented environmental policies.[6]

Within the United States, these changes in the politics of environmental policy represent a dramatic departure from long-term trends. Just a few years ago, the only serious consideration given to market-oriented environmental-protection policies was by economists at universities and research institutions. A new environmentalism has emerged that embraces these innovative approaches.[7] The Environmental Defense Fund was the first of the major environmental organizations to advocate incentive-based policies; there is growing support for these approaches by other groups as well.[8] Because of the important political role played by the major environmental advocacy groups in the United States, such support is of major consequence.

THE CONTEXT OF CHANGE IN THE UNITED STATES: ENVIRONMENTAL AND ECONOMIC CONDITIONS, PUBLIC OPINION, AND CURRENT POLICY

U.S. environmental and natural resource policies have evolved over the past twenty years in response to an array of perceived risks. New threats have arisen,[9] the cost of enforcing existing policies has escalated, and the issue of how to share that burden frequently has become a brake on needed action. It is less and less likely that we can increase environmental protec-

5. David Pearce, Anil Markandya, and Edward B. Barbier, *Blueprint for a Green Economy* (London: Earthscan, 1989).

6. In the Soviet Union, the Central Institute of Mathematics and Economics (Tsemi) of the Academy of Sciences has advocated the use of pollution taxes for a variety of environmental problems. In Poland, government officials have endorsed marketable-permit programs to help address air- and water-pollution problems (John Palmisano, personal communication, December 8, 1989).

7. See Frederic D. Krupp, "New Environmentalism Factors in Economic Needs," *The Wall Street Journal*, November 20, 1986, p. 34.

8. Among the leading environmental advocacy groups, the Wilderness Society, the National Audubon Society, the Sierra Club, and the Natural Resources Defense Council have come to support selective use of economic-incentive mechanisms. The recent endorsement by the Natural Resources Defense Council, a leading supporter of clean-air reforms, for a tradable-permit system for acid-rain control is particularly important. See Matthew L. Wald, "Searching for Incentives to Entice Polluters," *The New York Times*, October 8, 1989.

9. A discussion of future environmental challenges is provided by Milton Russell, "Environmental Protection for the 1990s and Beyond," *Environment* 29 (1987), pp. 12–38.

tion simply by spending more money on programs and policies already in place.[10] Moreover, the costs of environmental compliance to the economy, as a whole, continue to increase.[11] If the country is to build and maintain international competitive strength along with a better environment, it is important to ensure that investment in environmental protection is cost-effective. The approach that seems most promising involves harnessing market forces to spur both technological advance and sustainable management of natural resources.

By channeling the forces of the marketplace into environmental programs, economic-incentive mechanisms can make the everyday economic decisions of individuals, businesses, and government work effectively for the environment. This does not mean using economic criteria, exclusively or otherwise, to set environmental goals. There is no recommendation here of using cost/benefit analysis or setting dollar values on environmental amenities or human health. Indeed, in many cases ongoing debates over specific environmental standards should be set aside, so that separate examination of effective mechanisms for environmental protection can be carried out.

Developing such proposals in detail and putting them into action will be a complicated and difficult enterprise, but this is a challenge that must be met. Public demand in the United States for a quality environment has been strong and consistent.[12] Yet it is unlikely that either the federal government or the U.S. economy as a whole will be able to afford substantially higher environmental standards unless means are found that achieve the most protection possible for every dollar spent. The conventional "command and control" approach of setting uniform standards or requiring specific control technologies is an increasingly difficult and expensive method to achieve environmental improvements. Market forces can supplement the regulatory

10. Federal expenditures for all environmental and natural-resource programs in 1985 were about $13.4 billion (1.4 percent of all federal outlays). See U.S. Office of Management and Budget, *Budget of the U.S. Government, Historical Tables, Fiscal Year 1987* (Washington, DC: U.S. Government Printing Office, 1986); and U.S. Council of Economic Advisors, *Economic Report of the President.* (Washington, DC: U.S. Government Printing Office, 1986).

11. In 1984, total U.S. expenditures on pollution control amounted to about $65 billion—63 percent by businesses, 21 percent by all levels of government, and 16 percent by consumers. Total pollution control expenditures were about 1.8 percent of Gross National Product. See Kit D. Farber and Gary L. Rutledge, "Pollution Abatement and Control Expenditures," *Survey of Current Business* 66 (1986), pp. 100–103.

12. Public-opinion polls show consistently that public concern over environmental quality has remained firm during energy crises, economic downturns, and tax revolts. See Riley E. Dunlap, "Polls, Pollution, and Politics Revisited: Public Opinion on the Environment in the Reagan Era," *Environment* 29 (1987), pp. 7–37; E. C. Ladd, "Clearing the Air: Public Opinion and Public Policy on the Environment," *Public Opinion,* February/March 1982, pp. 16–20; and Richard D. Lamm, and Thomas A. Barron, "The Environmental Agenda for the Next Administration," *Environment* 30 (1988), pp. 17–29.

power of the government and create a setting for private-sector innovation and initiative in the pursuit of environmental quality.

CRITERIA FOR ENLIGHTENED PUBLIC POLICY: IDENTIFYING INNOVATIVE SOLUTIONS TO ENVIRONMENTAL PROBLEMS

In order to identify appropriate policies for specific environmental and natural-resource problems, a variety of questions need to be considered, among them:

- Will the policy achieve environmental goals effectively?
- Will the policy approach be cost-effective, that is, will it achieve environmental goals at a minimal cost to society at large?
- Will the strategy provide relevant government agencies with the information they need?
- How easy and costly will monitoring and enforcement be?
- Will the policy be flexible in the face of change? When changes occur in tastes, technology, or use of resources, will the policy accommodate these changes and remain effective, or will it become ineffective or even counterproductive?
- Will the policy give industry positive, dynamic incentives? Will it, for example, encourage firms to develop new, environment-saving technologies, or to retain existing, inefficient plants?
- Will the economic effects of the policy be equitably distributed?
- Will the purpose and nature of the policy be broadly understandable to the general public?
- Will the policy be truly feasible, in terms of enactment and implementation?

On the basis of the answers to these questions, effective new approaches to recognized problems can be developed.[13]

---

13. This set of criteria is based partly on a similar set of criteria described by Peter Bohm and Clifford S. Russell, "Comparative Analysis of Alternative Policy Instruments," in Allen V. Kneese and James L. Sweeney, eds., *Handbook of Natural Resource and Energy Economics* (Amsterdam: North-Holland, 1985), vol. 1, pp. 395–460.

ECONOMIC-INCENTIVE MECHANISMS: WHAT THEY ARE AND HOW THEY WORK

A key to reducing inefficient use of natural resources and environmental degradation is to ensure that consumers and producers face the true costs of their decisions—not just their direct costs but the full social costs and consequences of their actions. Economic-incentive systems provide various ways to do this. Most such approaches can be viewed as falling within one of five major categories: pollution charges, marketable permits, deposit-refund systems, removal of market barriers, and elimination of government subsidies.

Pollution Charges

Charge systems impose a fee or tax on pollution (*not* simply on pollution-generating activities). As a result, it pays businesses to reduce pollution (up to the point where the marginal cost of pollution control is equal to the pollution tax rate). Thus, companies wind up controlling different amounts of pollution (the high-cost controller controls less; the low-cost controller controls more), but all will tend to experience the same marginal cost of pollution control. The result is that the total costs of pollution control are minimized, as compared with other allocations of the pollution-control burden among businesses. Under this and other economic-incentive approaches, additional pollution-control efforts are in the financial interests of business, as long as the costs of controlling pollution are sufficiently low. These mechanisms provide ongoing incentives for firms to develop and adopt newer, better (and cheaper) pollution-control technologies.

Examples of water-pollution charges are found in several European nations, including France, the Netherlands, and West Germany. One frequently discussed application is a carbon tax to help control global warming.

One potential problem with emission-charge systems is that governments do not know in advance what level of cleanup will result from any given charge. Tradable-permit systems eliminate this particular problem.

Marketable-Permit Systems

Marketable or tradable permits can achieve the same cost-minimizing allocation of the pollution-control burden, but they do so in ways that avoid the aforementioned problem of uncertain business responses. Under a tradable-permits system, the allowable overall level of pollution is established and then allotted in the form of permits among companies. Those that keep emission levels below the allotted level may sell their surplus permits to other firms or use them to offset excess emissions in other parts of their own facilities. Low-cost controllers thus have an incentive to control more. High-cost controllers, who may buy permits instead of undertaking expensive control measures, thus control less. As with a charge system, the marginal cost of control is identical for all firms; the total, societal cost of control is minimized for any given level of total pollution control.

Neither taxes nor permit systems need begin or remain at the status quo level of pollution (or pollution control); permit systems, for example, can issue initial permits for some fraction of current emissions and establish deadlines for achievement. Likewise, they can be used to move toward stricter standards.

The primary application of such mechanisms has been in the United States, both under the Environmental Protection Agency's (EPA) Emissions Trading Program and the nationwide lead phasedown, which allowed fuel refiners to "bank" and "trade." Both the U.S. House and Senate recently adopted clean air legislation that will institute a marketable-permit system for acid-rain control. Other potential areas of application include: local air pollution, point- and nonpoint-source water pollution, chlorofluorocarbon (CFC) reduction, and control of global warming through international trading in greenhouse gas permits and offsets.

Deposit-Refund Systems

Under this approach, a deposit or surcharge is paid when a potentially polluting product or product-packaging is purchased. When the consumer returns the polluting materials to an approved center (for recycling or proper disposal), the deposit is refunded. This approach has been used successfully in a number of states with so-called bottle bills to reduce littering with beverage containers *and* to reduce the flow of solid waste to costly landfills. Deposit-refund systems can be used also for containerizable hazardous waste and for certain forms of solid waste. Lead-acid batteries, used motor-vehicle oil, and tires are obvious candidates.

One advantage of the deposit-refund system is that it eliminates the incentive for illegal "midnight dumping" that exists under a simple waste-end tax or fee. There are proposals in the Congress for applying the deposit-refund concept to new problem areas. Our nation might adopt the practices of other countries: Denmark, for instance, has a deposit-refund plan for mercury and cadmium batteries, and Norway and Sweden have successful systems for car hulks.

Removing Market Barriers

In some cases, substantial gains can be made in environmental protection simply by removing government-mandated barriers to market activity. For example, measures that facilitate the voluntary exchange of water rights can promote more efficient allocation and use of scarce water supplies, while curbing the need for expensive and environmentally disruptive new water-supply projects. Negotiations are under way for a major market-oriented water exchange in southern California. Other applications of the general concept include comprehensive least-cost bidding at electrical utilities, a measure that would promote economically rational energy conservation.

Eliminating Government Subsidies

Several existing subsidies promote inefficient and environmentally unsound development. A major example is in below-cost timber sales, by which the U.S. Forest Service does not recover the full cost of making timber available. The result of these subsidies has been excessive timber cutting, which leads to substantial loss of habitats and damage to watershed values. Gradual removal of subsidies would foster environmental protection and increase net federal revenues.

Other examples of subsidies that are both economically inefficient and environmentally disruptive include those associated with U.S. Army Corps of Engineers flood-control projects, U.S. Bureau of Reclamation projects, and agricultural price supports.

Comparing Market-Based Mechanisms with Conventional Policies

At a time of concern about international competitiveness, incentive-based approaches can provide huge savings and increases in productivity. For example, a market-based approach to acid-rain reduction could save up to $3 billion per year, compared with the cost of a dictated technological solution.[14] And incentive-based approaches need not be any more expensive for the government to administer than conventional regulatory methods. In fact, funds from tradable-permit auctions could help finance an expanded EPA budget.[15] Such systems provide incentives for one firm to monitor the activities of other pollutant-emitting firms—another manifestation of the discipline of a competitive market. This is not to suggest, however, that environmental protection can be achieved without significant government expenditures; no program of controls can be effective without a strong commitment by government to monitoring and enforcement.

Most important, economic-incentive approaches allow greater levels of protection for any given aggregate cost of control. Rather than dictate to enterprises how they should manufacture their products, incentive-based systems impose a cost on pollution-causing activities, leaving it to individual firms to decide among themselves how to achieve the required level of environmental protection. Market forces will drive these decisions toward least-cost solutions and toward the development of new pollution-control technologies and expertise by the private sector.

Market forces also lead to powerful incentives for the development of new pollution-control technologies and expertise by the private sector. Because investments in pollution control lead to tangible, positive effects on profits under incentive-based systems, these policies can provide significant inducements for firms to adopt pollution-control technologies. In turn, incentives are created for those same firms or others to carry out research and development of cheaper and better pollution-abatement techniques.[16]

Incentive-based approaches have an added benefit: they can make the environmental debate more understandable to the general public. Because they do not dictate a particular technology, these approaches can focus

14. ICF Resources, *Analysis of Six and Eight Million Ton 30-Year/NSPS and 30-Year/1.2 Pound Sulphur Dioxide Emission Reduction Cases* (Washington, DC, February 1986). Recent estimates indicate that the current administration's proposed marketable-permit program for acid-rain control would save between $13 and $16 billion by the year 2010, compared with a conventional approach. See Robert W. Hahn, "Designing Markets in Tradable Allowances for Reducing Acid Deposition," draft manuscript (Washington, DC, December 1989).

15. For further discussion of such possibilities, see Bruce A. Ackerman and Richard B. Stewart, "Reforming Environmental Law: The Democratic Case for Market Incentives," *Columbia Journal of Environmental Law* 13 (1988), pp. 171–199.

16. In general, the relative superiority (in terms of inducing technological innovation and diffusion) of incentive-based approaches, compared with conventional command-and-control approaches, is clear. See Scott R. Milliman and Raymond Prince, "Firm Incentives to Promote Technological Change in Pollution Control," *Journal of Environmental Economics and Management* 17 (1989), pp. 247–265. Under certain circumstances, however, emission-credit trading may reduce corporate incentives to adopt new technology. On this see David A. Malueg, "Emission Credit Trading and the Incentive to Adopt New Pollution Abatement Technology," *Journal of Environmental Economics and Management* 16 (1989), pp. 52–57.

attention directly on what our environmental goals should be, rather than on difficult technical questions concerning technological alternatives for reaching those goals.

Market-oriented policies, however, will not fit every problem. On the one hand, incentive-based approaches are virtually tailor-made for problems such as acid rain, for which concern focuses on *aggregate* pollution levels within an airshed. As explained above, economic-incentive mechanisms such as tradable permits allocate the pollution burden across businesses in such a way as to minimize total expenditures for any given level of aggregate control. On the other hand, for environmental problems that display local *and* threshold effects, concern focuses on the level of pollution emitted by individual sources. In these cases, a conventional command-and-control approach (e.g., a uniform emission standard) may be the preferred policy.[17]

The best set of policies probably will involve a mix of market and more conventional regulatory processes. To design and implement economic-incentive programs, it will be necessary to adapt, not abandon, present programs and build step by step with market-based policies on the initiatives of the United States and other industrialized nations.

PREVIOUS U.S. EXPERIENCE WITH INCENTIVE-BASED ENVIRONMENTAL POLICIES

As noted above, market-based approaches for environmental protection have been implemented on a limited scale in the United States and several European nations. Three of our country's experiences are described in more detail below.

The EPA's Emissions-Trading Program

In 1974, the Environmental Protection Agency began to experiment with "emissions trading" as part of its program for the improvement of local air

---

17. Even in the former case, where concern focuses on aggregate pollution levels, there are situations in which command-and-control approaches may turn out to be more efficient than incentive-based approaches (*if* the two approaches accomplish different levels of pollution control). See Wallace E. Oates, Paul A. Portney, and Albert M. McGartland, "The Net Benefits of Incentive-Based Regulation: A Case Study of Environmental Standard Setting," *American Economic Review* 79 (1989), pp. 1233–1242.

quality.[18] Businesses that reduce emissions below the level required by law have been allowed to receive "credits" that can be applied against higher emissions elsewhere. Under programs of "netting" and "bubbles," firms have been permitted to trade emissions reductions among sources within each firm, so long as total, combined emissions comply with an aggregate limit.

Businesses have also traded emissions credits. Under the "offset" program, begun in 1976, firms that wish to establish new sources in areas not in compliance with ambient standards have been required to offset their new emissions by reducing existing emissions by a greater amount. This can be done with their own sources or through agreements with other companies. Finally, under the "banking" program, businesses may store earned emissions credits for future use, to allow either for internal expansion or for a sale of credits to other firms.

These programs were codified in the EPA's Final Policy Statement on Emissions Trading in 1986, but their use to date has not been extensive. States are not required to use them, and uncertainties about the future course of the programs have made companies reluctant to participate. Nevertheless, Armco, Du Pont, USX, and 3M, among others, have traded their emissions credits, and a limited market for transfers has arisen. Even this degree of participation in the EPA's programs has resulted in more than $4 billion in savings in control costs, with no adverse effect on air quality.

Tradable Permits for Water-Pollution Control

Nonpoint sources, particularly from agriculture and urban runoff, now constitute the major American water-pollution problem. The experience of Dillon Reservoir, the major source of water for the city of Denver, provides an example of a trading approach for nonpoint-source water pollution. In past years, nitrogen and phosphorus loading was turning the reservoir eutrophic (rich in dissolved nutrients but often shallow and seasonally deficient in oxygen), despite the fact that point sources from surrounding

---

18. For a recent description of the EPA's (and some European) experiences with incentive-based policies, see Robert W. Hahn, "Economic Prescriptions for Environmental Problems: How the Patient Followed the Doctor's Orders," *Journal of Economic Perspectives* 3 (1989), pp. 95–114. A different perspective is provided by Michael H. Levin, "New Directions in Environmental Policy: The Case for Environmental Incentives," *Proceedings of Annual Midwinter Meeting, American Bar Association, Section of Natural Resource Law* (Keystone, Colo., March 18–20, 1988).

communities were controlled to best-available-technology standards. In order to preserve and protect water quality in the face of rapid population growth, a "point/nonpoint-source control optimization" program was developed to cut phosphorus flows mainly from nonpoint urban and agricultural sources.

The point/nonpoint-source trading plan was developed with active participation of environmental groups, industry, and local and state governments, and was approved by Colorado and the EPA in 1984. The program allows for publicly owned sewage treatment works (POTWs) to finance the control of nonpoint sources in lieu of upgrading their own treated effluent to drinking-water standards. The program is effective because the cost per pound of phosphorus removed via trading is $67, versus $824 per pound for the cheapest advanced treatment alternative developed for the POTWs. The EPA has estimated that the plan has made aggregate savings of over $1 million per year, compared with the conventional workings of four fairly small POTWs.

Voluntary Water Exchanges

One effective approach to water-supply problems is to allow the voluntary exchange of water rights in order to increase efficiency—most notably by creating economic incentives for water conservation. In the Imperial Irrigation District (IID) of California, farmers are paying as little as $10 for water to irrigate an acre of cotton, while just a few hundred miles away in Los Angeles, local authorities of the Metropolitan Water District (MWD) are paying up to $200 for the same quantity of water. A free market in water rights, allowing voluntary exchanges, would make both parties better off: farmers would have a financial stake in conserving water, urban needs would be met without shrinking agriculture and without building new dams and reservoirs, and environmental protection would gain.

In March 1983, the Environmental Defense Fund (EDF) published a proposal calling for the MWD to finance the modernization of the IID's water system in exchange for use of conserved water. In November 1988, after five years of negotiation, the two water giants reached agreement on a historic $230 million water conservation and transfer arrangement,[19]

19. Willy Morris, "IID Approves State's First Water Swap with MWD," *Imperial Valley Press,* November 9, 1988.

which closely parallels the EDF's original proposal of "Trading Conservation Investments for Water."

This southern California water swap may be the harbinger of more enlightened western water policy, since it demonstrates that such trades can be executed on a significant scale. Such optimism seems to be validated by reports of greatly increased interest in water marketing in Colorado, New Mexico, Arizona, Nevada, Utah, and California.[20]

Other Examples

Although it is only during the past year that increased attention on a national level has been given to incentive-based environmental-protection strategies, additional examples exist of policies that are already operative. Among these are the EPA's tradable-permit system for implementing the Montreal Protocol's stratospheric ozone-depletion restrictions, the nation-wide lead phasedown, and some "experimental" use of comprehensive least-cost bidding by electric utilities.[21] Also, as noted above, several European nations have had significant experiences with economic-incentive mechanisms.[22]

POLICY RECOMMENDATIONS FOR SUSTAINABLE DEVELOPMENT

On the basis of the criteria described above and the experiences that the United States and other nations have already had with innovative incentive-based policies, it is possible to suggest specific mechanisms for individual problem areas. In the following sections, policy recommendations are offered for thirteen major environmental and natural resource problems faced by the United States:

1. The greenhouse effect and global climate change;
2. Stratospheric ozone depletion;

20. Sandra D. Atchison, "Where Water Is Money in the Bank," *Business Week*, August 15, 1988, p. 50.

21. Maine held one such auction earlier this year; Massachusetts and New York have announced their intentions to hold similar auctions in the near future.

22. See J. B. Opschoor, and Hans B. Vos, *The Application of Economic Instruments for Environmental Protection in OECD Countries*, Final Report of the Environment Committee (Paris: Organization for Economic Cooperation and Development, 1988). A brief overview is provided by James David Spellman, "Environmental Needs Challenge the Global Marketplace," *Europe Magazine*, September 1989, pp. 18–20.

3. Local air pollution;
4. Acid rain;
5. Indoor radon pollution;
6. Threats to energy security and environmental quality;
7. Inefficient use and allocation of water supplies;
8. Degradation of surface water and groundwater quality;
9. Solid-waste management;
10. Presence of toxic substances in the environment;
11. Management of toxic and infectious waste;
12. Public-lands management;
13. Depletion of wetland resources.

In each case, the problem is briefly defined and one or more promising policy measures is described.[23]

The Greenhouse Effect and Global Climate Change

The single most important environmental threat our planet has faced since the beginning of the Industrial Revolution may be global climate change due to the greenhouse effect, associated with emissions of carbon dioxide, methane, oxides of nitrogen and sulphur, and chlorofluorocarbons (CFCs). Consequent global temperature increases are predicted to produce climate changes unprecedented for their speed of occurrence, including massive changes in global precipitation patterns, storm intensities, and ocean levels.[24]

Recommendation: Fund research on causes and consequences of global warming and on specific adaptation and prevention strategies. Increased support of basic scientific research on the greenhouse effect and climate change is needed. The long-term goal should be to identify the most effective, efficient, and politically viable strategies—whether they are based on prevention, adaptation, or more likely, some combination of the two approaches.

Recommendation: Modify current policies to encourage energy efficiency. Encouraging efficient use of energy resources can reduce the burning of

23. The Project 88 study of Senators Heinz and Wirth offered thirty-six policy recommendations for these environmental and natural resource problem areas. A sample of thirteen recommendations drawn largely from that study are briefly summarized here.

24. Daniel J. Dudek, "Assessing the Implications of Changes in Carbon Dioxide Concentrations and Climate for Agriculture in the United States," *Preparing for Climate Change*, Proceedings of the First North American Conference on Preparing for Climate Change: A Cooperative Approach (Government Institutes, April 1988), pp. 428–450.

fossil fuels for electricity and increase energy efficiency in transportation and other sectors of the economy. One example of a policy change to consider would factor social—especially environmental—costs into the calculations used in the Public Utilities Regulatory Policy Act of 1978.

Recommendation: Prevent deforestation through "debt-for-forest" swaps. Common interests among developed and less developed nations could be furthered by extending the concept of offsets into the international arena through so-called debt-for-forest swaps, several of which have already been arranged by the World Wildlife Fund and other organizations.[25] Maintaining, rather than burning, tropical forests can constitute a significant source of greenhouse gas reductions; and voluntary debt-for-forest swaps will benefit debt-burdened LDCs.[26]

Stratospheric Ozone Depletion

Anthropogenic emissions of CFCs and other potential ozone-depleting chemicals are causing the depletion of stratospheric ozone. As a result, increased ultraviolet radiation will reach the Earth's surface, potentially elevating the incidence of skin cancer, promoting cataracts, suppressing immune responses, and causing other adverse effects to animals, plants, and valuable materials.

Recommendation: Phase out potential ozone depletors with the tradable permit system. To implement the Montreal Protocol's restrictions, the EPA has promulgated a system of nationwide tradable permits to achieve specified control levels much more economically and less disruptively than direct bans or controls. This market-based approach will stimulate firms to adopt measures tailored to specific circumstances, and will provide industry with incentives to develop substitute chemicals, industrial processes, and consumer products.[27]

25. Developing and developed nations have cooperated, to a limited degree, on international approaches to deforestation issues. An action plan has been cosponsored by the World Bank, the United Nations Development Programme, the Food and Agricultural Organization, and other agencies. See World Resources Institute, *Tropical Forests: A Call for Action* (Washington, DC: World Resources Institute, 1985).

26. Debt-for-forest swaps carried out between the U.S. government (directly or through commercial banks) and LDCs can produce intended, beneficial environmental effects only if LDCs are able to monitor and enforce the local execution of forest-saving plans. Related local administrative costs should therefore be included in designs of debt-for-forest swaps.

27. J. K. Hammitt, *Timing Regulations to Prevent Stratospheric-Ozone Depletion* (Santa Monica, CA: RAND, April 1987); A. S. Miller and I. M. Mintzer, *The Sky Is the Limit: Strategies for Protecting the Ozone Layer*, Research Report No. 3 (Washington, DC: World Resources Institute, November 1986).

Local Air Pollution

As a result of twenty years of federal attention to local air-pollution problems, there have been substantial improvements in air quality in most parts of the country. Nevertheless, more than 100 million Americans remain exposed to excessive smog (ambient ozone) levels, and some seventy urban areas still lack adequate local plans to reduce them.

Recommendation: Implement tradable permits for stationary sources. The EPA's initiatives with emissions trading already have saved the country an estimated $5 billion in costs for air-pollution control over the last several years, with equal or better environmental progress, according to the U.S. Government Accounting Office.[28] A logical extension is a comprehensive system of marketable emissions permits.[29] Sources that can reduce discharges below their permit levels would be allowed to sell part of their permitted surplus to others; firms for which compliance is relatively costly could control to the point at which it makes more sense to buy additional permits instead. Under this approach, sources will have continuous financial reasons to reduce beyond (not merely to) required levels. Air-quality goals can be achieved at lower cost overall.[30]

Acid Rain

The environmental consequences of acid rain appear to be escalating, but with the exception of Clean Coal Technology and related programs, Congress has been unable to enact explicit and effective policy responses. Many factors lie behind this failure, the major one being cost consideration and attendant social disruption, especially for communities in Appalachia and the Midwest that mine and burn high-sulphur coal.

Recommendation: Initiate an Acid-Rain Reduction Credit program.[31]

28. A critical examination of the EPA's experience is provided by Richard A. Liroff, *Reforming Air Pollution Regulations: The Toil and Trouble of EPA's Bubble* (Washington, DC: The Conservation Foundation, 1986). Some of the shortcomings in EPA's emission trading rules have been remedied; on this point, see Daniel J. Dudek and John Palmisano, "Emissions Trading: Why is This Thoroughbred Hobbled?" *Columbia Journal of Environmental Law* 13 (1988), pp. 217–256.

29. Robert W. Hahn, "Innovative Approaches for Revising the Clean Air Act," *Natural Resources Journal* 28 (1988), pp. 171–188.

30. Some environmental groups continue to express concerns regarding emissions trading and similar incentive-based approaches. For a comprehensive response, see Richard B. Stewart, "Controlling Environmental Risks Through Economic Incentives," *Columbia Journal of Environmental Law* 13 (1988), pp. 153–169.

31. This recommendation is embodied in the Bush administration's Clean Air Act amendment proposals, now under consideration by Congress.

Patterned after the EPA's emissions-trading program,[32] this economic-incentive approach to acid-rain control offers a cost-effective and equitable solution to this seemingly intractable problem. Any sources of emissions contributing to acid rain could use or sell "excess" reductions (beyond those required by law) to meet further emission-reduction requirements applicable to other sources. With this approach—which would complement, not replace, the traditional approach—goals for acid-rain reduction could be met at much lower cost than would otherwise be possible, with savings of up to 50 percent annually, according to the EPA.[33] If initial permits are auctioned, revenues can be used to finance the installation of retrofit and clean-coal technology through federal cost-sharing arrangements (in areas that currently utilize high-sulphur coal) and to support research-and-development efforts on the same technologies. The efficiency properties of the program thus can be combined with equitable protection for communities whose economies depend on the high-sulphur coal industry.

Indoor Radon Pollution

The EPA has identified radon as one of the most serious environmental risks facing the nation, causing 5,000 to 20,000 deaths from lung cancer each year. Because radon exposure occurs largely in private homes, it has not been feasible to use the conventional regulatory approach of setting and enforcing health-based exposure standards.

Recommendation: Give consideration to a variety of federal actions. Consideration should be given to a number of possible approaches, including tax incentives and subsidized loans, construction codes, soil tests, testing requirements for real estate transactions, and accelerated dissemination of information.[34]

32. Tom Tietenberg, *Emissions Trading: An Exercise in Reforming Pollution Policy.* Washington, DC: Resources of the Future, 1985.

33. ICF Resources, *Economic Analysis of Title V (Acid Rain Provisions) of the Administration's Proposed Clean Air Act Amendments (H.R. 3030/S. 1490),* prepared for the U.S. Environmental Protection Agency (Washington, DC: ICF Resources, September 1989).

34. F. Reed Johnson and Ralph A. Luken, "Radon Risk Information and Voluntary Protection: Evidence from a Natural Experiment," *Risk Analysis* 7 (1987), pp. 97–107.

Threats to Energy Security and
Environmental Quality

Crude petroleum accounts for more than 40 percent of the nation's energy needs; during the past twenty years, an increasing share of the crude oil used in this country has been provided by imports. Attempts to obtain domestic energy security through higher production of fossil fuels will run up against serious environmental problems. Instead, the United States can make significant gains in energy security and environmental protection by pursuing strategies of increased energy efficiency.

Recommendation: Increase energy efficiency through comprehensive least-cost bidding at electrical utilities. Opening U.S. power markets and allowing methods of demand management to compete head to head with power production is a market-oriented approach that some states have already taken. Systems of comprehensive least-cost bidding will increase the efficiency of energy production substantially.[35]

Inefficient Use and Allocation of
Water Supplies

If current practices continue, water shortages may become commonplace in the United States during the next two decades. Policies at present do not give users appropriate incentives to take actions consistent with economic and environmental values of water resources. The lack of such incentives results in grossly inefficient use of existing supplies, since decision makers do not experience the true costs of their decisions on water use.[36]

Recommendation: Remove barriers to water markets. Measures that facilitate voluntary water transfers will promote more efficient allocation and use of scarce water resources and curb the need for more (expensive and environmentally disruptive) water-supply projects.[37] The government should begin to remove barriers to voluntary water marketing by validating

35. Charles Cicchetti and William Hogan, *Including Unbundled Demand Side Options in Electric Utility Bidding Programs,* Energy and Environmental Policy Center Discussion Paper E-88-07 (Cambridge, Mass.: Harvard, August 1988).

36. A timely and comprehensive study of federal water projects and the potential for market-oriented reforms is Richard W. Wahl, *Markets for Federal Water: Subsidies, Property Rights, and the Bureau of Reclamation* (Washington, DC: Resources for the Future, 1989).

37. Robert N. Stavins and Zach Willey, "Trading Conservation Investments for Water," in R. J. Charbenean, ed., *Regional and State Water Resources Planning and Management* (Bethesda, Md.: American Water Resources Association, 1983), pp. 223–230.

voluntary transfers of water and establishing rules to protect public and other third-party uses of water.[38]

Degradation of Surface Water and Groundwater Quality

Most laws and regulations regarding water-pollution control in the United States have been directed exclusively at point sources, such as factories and municipal waste-treatment facilities. Dispersed, nonpoint sources—including farms and urban runoff—have not been adequately addressed, in part because such sources are much more difficult to control, particularly by conventional methods. An increasingly serious problem is the contamination of groundwater supplies by seepage of hazardous chemicals stored in dumpsites and municipal landfills, leaks from underground storage tanks, highway runoff, and infiltration of pesticides and fertilizer residues.

Recommendation: Focus the U.S. Department of Agriculture's Conservation Reserve Program (CRP) on water-quality problems. This would be an important step toward control of nonpoint-source water pollution from agricultural runoff. Eligibility criteria for the CRP should be broadened to include lands that are important in terms of water pollution. States and private trusts should be encouraged to pay bonuses above federal contributions to attract into the reserve lands that meet state water-quality goals.[39]

Recommendation: Implement tradable discharge permits for point sources. Existing approaches to point-source control, while holding each source to specified limits, do not restrain the total volume of discharges within a basin. Where this a concern, the establishment of an overall watershed limit and the implementation of tradable permits within it will be the most effective and efficient way of achieving water-quality goals.

Solid-Waste Management

Throughout the country, old landfills are reaching the saturation point, and it is becoming more and more difficult to find sites for new ones. Giant garbage incinerators are bringing with them equally giant bond issues

38. Zach Willey and Tom Graff, "Federal Water Policy in the United States—An Agenda for Economic and Environmental Reform," *Columbia Journal of Environmental Law* 13 (1988), pp. 325–356.

39. Tim Phipps, "The Conservation Reserve: A One Year Progress Report," *Resources,* Winter 1987.

representing burdensome investments for many communities; and now it is becoming clear that incinerators produce their own set of significant environmental hazards.

Recommendation: Implement policies that allow recycling to compete. The vast majority of our garbage is recyclable. The critical question is whether recycling makes economic sense, and the answer is that the most important economic benefits of recycling typically come from reducing the quantity of garbage that otherwise must be collected and disposed of, not from revenue due to sales of recycled materials.[40] If communities are to adopt efficient solutions to their problems of solid-waste management, recycling must be considered on an equal basis with other alternatives. The bidding process for municipal waste management should be opened to all techniques, by specifying outputs and results rather than specific techniques.

*Presence of Toxic Substances
in the Environment*

Although federal legislation nominally encourages the reduction of toxic-waste sources, the actual focus of regulation has been on controlling pollution at the "end of the pipe."

Recommendation: Provide incentives for source reduction. To finance the cleanup of hazardous-waste sites, the Superfund program imposes a front-end tax on the chemical and petroleum industries, unrelated to the toxicity of products or services. This tax provides no incentive for firms to switch to less hazardous substances or to recycle wastes. A waste-end tax might induce industries to reduce the toxicity of their products, packaging, and processes and might provide an incentive to consumers to use safer seals. But waste-end taxes provide incentives for illegal dumping. In some cases, the answer to this quandary will be a deposit-refund system, discussed below. Another approach to source reduction is through labeling requirements, which compel producers to inform consumers that products or product-packaging contain known toxic substances that may present significant risks.[41] This approach must be used only in limited cases, however,

---

40. Environmental Defense Fund, *Coming Full Circle: Successful Recycling Today* (New York: Environmental Defense Fund, 1988).

41. For alternative perspectives on the use of labeling, see David Roe, "Barking Up the Right Tree: Recent Progress in Focusing the Toxics Issue," *Columbia Journal of Environmental Law* 13 (1988), pp. 275–283; and Melinda Haag, "Proposition 65's Right-To-Know Provision: Can It Keep Its Promise to California Voters?" *Ecology Law Quarterly* 14 (1987), pp. 685–712.

since excessive labeling may simply cause people to ignore signs or labels that warn of genuine risks.

Management of Toxic and Infectious Waste

Among the most difficult of hazardous-waste problems are those posed by wastes generated in quantities small enough that they can be stored, shipped, and dumped more or less anywhere in the environment.

Recommendation: Implement a deposit-refund system for containerized wastes. A front-end tax, which works as a deposit, with a refund payable when quantities of toxic substances are turned in to designated facilities for recycling or disposal, would provide important incentives for toxic-waste management.[42] First, there is an incentive to follow rules for proper disposal and to recapture would-be losses from the production process; second, there is an incentive for producers to look for nonhazardous substitutes; and third, agencies' monitoring problems will no longer include the nearly impossible task of preventing illegal dumping of small quantities at dispersed sites in the environment.

Public-Lands Management

The federal lands of the United States contain valuable natural resources— timber, minerals, oil, natural gas, forage for livestock, and so on—and hold a precious endowment of environmental resources, including wilderness, fish and wildlife and their habitats, watershed values, free-flowing rivers and streams, scenic beauty, outdoor recreational opportunities, and untapped scientific information. At the least, public lands should be managed to provide benefits that private lands are unlikely to produce in our market economy. Sound management of these public lands is seriously impeded, however, because costly subsidies exist for a few extractive industries at the expense of the public's interests in environmental values and outdoor recreation.

Recommendation: Reduce government subsidies. The largest subsidy, and the one most in conflict with environmental values, is that given to timber

42. Clifford S. Russell, "Economic Incentives in the Management of Hazardous Wastes," *Columbia Journal of Environmental Law* 13 (1988), pp. 257–274. An earlier and more general inquiry is provided by Peter Bohm, *Deposit-Refund Systems: Theory and Applications to Environmental, Conservation, and Consumer Policy* (Washington, DC: Resources for the Future, 1981).

sales in remote, roadless areas of the national forests. Low-value timber is frequently sold in environmentally and recreationally valuable areas where roadbuilding to reach and harvest timber is expensive and damaging.[43] Below-cost timber sales—for which the U.S. Forest Service does not recover the full cost of making timber available—have cost the U.S. Treasury more than $400 million annually over the past six years. Gradual removal of the subsidies would foster protection of the environment, decrease federal expenditures, and increase net revenues.

Depletion of Wetland Resources

For 200 years, wetlands have been drained, cleared, and filled for agricultural, municipal, and industrial uses. In their natural state, however, wetlands also produce significant benefits—regulation of water flows, filtration and purification of water, and provision of habitat for flora and fauna. Wetland losses in the United States now average 450,000 acres annually, an area almost half the size of Rhode Island, with agriculture accounting for nearly 90 percent of recent conversions.

Recommendation: Improve use of environmental-impact statements. Federal flood-control and drainage projects provide financial incentives for private landowners to convert their wetland holdings to dry croplands.[44] Such impacts should be candidly assessed through the process of the Environmental Impact Statement. Impact areas must be correctly defined to include areas where drainage and clearing are (economically) induced.

Recommendation: Institute market incentives to reflect wetland values. Government subsidies that promote economically inefficient and environmentally unsound development should be removed. Among the policy initiatives to consider are ending totally subsidized construction of federal flood-control and drainage projects,[45] eliminating favorable tax treatment of wetland conversion, and implementing cross-compliance legislation linked to receipt of federal commodity program payments.

43. Michael D. Bowes, and John V. Krutilla, *Multiple-Use Management: The Economics of Public Forestlands* (Washington, DC: Resources for the Future, 1989).

44. Robert N. Stavins and Adam B. Jaffe, "Unintended Impacts of Public Investments on Private Decisions: The Depletion of Forested Wetlands," *American Economic Review* 80 (1990), on press.

45. Robert N. Stavins, "Alternative Renewable Resource Strategies: A Simulation of Optimal Use," *Journal of Environmental Economics and Management* 18 (1990), on press.

MARKET-BASED ENVIRONMENTALISM: A NEW ERA FROM AN OLD IDEA?

Across the United States and throughout the industrialized world, there continues to be a strong consensus in favor of effective environmental protection. In many cases, environmental goals are clear—the question is how to achieve those goals. The policy tools chosen *do* make a difference.

Although conventional regulatory policies have sometimes worked well, they have tended to pit economic and environmental goals against each other, too often producing paralysis rather than progress. But in the long run, economic and environmental goals must complement one another, if either set of goals is to be achieved. This dichotomy can be addressed by applying economic-incentive mechanisms to the work of environmental protection.

In fairness to future generations, it is essential to begin now to deal with long-term economic and environmental problems. Sustainable solutions to today's problems are required, because the debts we incur now, whether economic or environmental, will have to be paid someday.

ROBERT ORNSTEIN

Why We Don't Listen . . .

Robert Ornstein is president of the Institute for the Study of Human Knowledge and teaches at the University of California Medical Center in San Francisco and at Stanford University. He has researched the human brain extensively and has written or co-authored many books, including *The Psychology of Consciousness, Multimind,* and most recently, *New World/New Mind* (with Paul Ehrlich). Ornstein has been director of the Fund for Conservation Biology since 1983.

To understand how and why we human beings have difficulty in changing our actions is a little beyond my short essay, yet an awareness of how the mind is structured may help a bit. It is not only the greenhouse problem that eludes our solution, but many other modern predicaments.

Why does the growing greenhouse effect and its related problems attract little attention, while a stock market "crash" makes headlines? we might ask. But we might also ask: Why does the number of nuclear weapons expand astronomically, though largely unheralded, while one hostage's plight commands the front pages? Why do we collectively spend billions on medical care while neglecting the simple preventive actions which, if we took them, would save many times that amount and benefit society much more?

All these things are happening now, and are happening all at once, in part because *the human mental system is failing to comprehend the modern world.*

We are all collectively in the same boat. We hear of the problems of getting ideas into something called "public opinion." We forget that public opinion, the media, the government are all composed of individual human beings with a nervous system and set of judgment priorities that make them *ignore* constant problems. We are blind and deaf to the real world.

Our brains are designed to respond to sharp change in the world, not slow and gradual change. And it is the slow changes that lie behind our current problems: stockpiles of nuclear weapons grow larger, budget deficits mount, our education becomes more and more obsolete, and the environment deteriorates. But most people's attention is fixed on eye-catching "images"—the little girl Jessica, who fell down a well, for example, or horrible murders, airplane crashes, dramatic changes in stock prices, surprising football scores. Cancer terrifies us, and we keep on smoking.

In the old world, for which our perceptual systems were designed, the overall environment remained relatively stable, limited. Threats were signaled by short-term changes, usually requiring immediate action. Humanity's ancestors faced these threats over millions of years of evolutionary history. Apes, australopithecines (our first upright ancestors), early human hunters and gatherers, and the inhabitants of early civilizations, like other animals, had evolved quick reflexes to deal adequately with such threats.

The world that made us is now gone, and the world we have made is a new one that we have developed little capacity to comprehend.

There now exists a mismatch between the human mind and the world humans inhabit. We have made radical transformations in the world in an instant, in evolutionary time. The mismatch interferes with the relationships of human beings to each other and with their environments. Human beings, and all other organisms, have to adapt to the environments in which they live. For most of the history of life our prehuman ancestors evolved biologically, as do all living things. Then, for the relatively brief period of human history—a few million years—adaptation took place primarily by means of cultural change: the development of language and tools; the invention of agriculture, cities, industry, and high technology.

Cultural evolution can be much more rapid than biological evolution, for it involves alterations of information stored in minds or in books, tools, art, and other artifacts of societies. Cultural evolution can effect significant changes in a matter of decades or less. But the rapid changes human beings are making in the world now have rendered even the pace of most cultural evolution far too slow.

We are losing control of our future. The serious and dangerous mismatch is this: Our civilization is threatened today by changes taking place over periods of years and decades, but changes over a few years or decades are too *slow* for us to perceive readily. That is a time scale too leisurely for a nervous system attuned to bears, branches, burglars, and downpours. At the same time, these changes are much too *rapid* to allow processes of biological or cultural evolution to adapt people to them. *We are out of joint with the times.* Our times.

And the rate of change in the world around us is increasing. Now humanity is refashioning the world so quickly that each *decade's* environment differs dramatically from that of the last. Each triumph of technology contains new threats. With the advent of television and other modern communications, we can even feel threatened by events (such as earthquakes in Soviet Armenia or San Francisco) occurring thousands of miles away.

The physiological tendency is to respond immediately to these events as if they were local emergencies, while at the same time we ignore things that happen to us or our neighbors that really *are* serious threats. Thus our old mental system struggles and often fails to distinguish the relevant from the trivial, the local from the distant, just as the ability to make such distinctions becomes increasingly crucial.

Consider the way we deal with common problems. Crack is the subject of much attention in the media, with its drug czar, William Bennett. But suppose I told you about a drug six times as addictive as crack that will kill approximately 4 million Americans in the 1990s, far more than crack will. One might feel it is imperative for us to act. But the drug is tobacco, and it has been difficult to whip up truly affirmative programs of action, even in view of the threat. New dramatic threats, then, receive attention; old ones form a part of the woodwork. Slow and familiar threats, the threats that are killing us, are ignored.

The human mental hardware—our senses and brains—is effectively fixed, stationary. Although we are evolving, our mental machinery will not change biologically in sufficient time to help us solve our problems, because the same mental routines that originally developed to signal abrupt physical changes in the old world are now pressed into service to perceive and decide about unprecedented dangers in the new.

Crack fits into the mind's routines, cigarettes do not. A tragic earthquake commands front-page news all over the world, even though twice as many people were killed on the highways that same weekend as in the earthquake, and each weekend since. We caricature the world.

A caricature simplifies reality so that much of the environment is not registered on an organism's sensory system. Like a political cartoonist's caricature of a president's face, only a few aspects of reality are emphasized. Why should that matter? It matters because for billions of years of evolution, our ancestors acted in situations in which extreme caricatures sufficed for survival. In order to understand our present limitations, we have to understand their origins. Evolution is frugal; it would never favor organisms that invested energy in sensory frills if that same energy could be used to enhance reproduction.

A limited ability to sense the environment has been built into all animals through eons of natural selection. *All* sensory systems filter information from the outside world, their environment. The human sensory system is no exception; we too live in a world of caricatures. We, for example, are unable to see patterns of ultraviolet light quite visible to butterflies searching for nectar. People spend little time sipping nectar from flowers, and so evolution has not provided us with the capacity to see the ultraviolet designs on petals that guide insects to a sugary delight. We cannot hear the sound of a dog whistle or, like a bloodhound, smell the scent of an escaped prisoner. And we can't perceive certain novel hazards that exist today. The radiation from Chernobyl is real enough to kill, but we cannot immediately sense or feel its insidious effects.

This unconscious cultural evolution developed in a small-group animal with a short time horizon, the human being, in possession of its culture. Humanity is inadequate to deal with a world overpopulated with individuals who are only partially in contact with their own cultures and who must make critical decisions about the medium- and long-term future.

Unconscious cultural evolution has not led humanity to pay explicit attention to biological or cultural evolutionary heritage. Cultural evolution has not compensated for the baggage of an outdated human perceptual system. It has not, for example, invented a "time-lapse" system for perceiving the gradual changes to Earth's atmospheric composition that human biological systems are incapable of sensing. It has not led school curricula to convey the limits of the human perceptual system. It has not led to the establishment of governmental institutions that force politicians to address and account for the long-term consequences of their actions. It has not generated television programs designed to produce a widespread awareness of the diverse limitations and built-in biases imposed on people by their biological and cultural evolutionary history. It has not provided us with an inventory of tools specifically designed to overcome biases. Cultural evolution has not even allowed most human beings to perceive that their familiar world results from an ongoing evolutionary *process,* even as it accelerated that process to unprecedented rates of change. It has not, therefore, given us the means of survival.

A single accident, the tragic and preventable loss of seven lives in the Challenger space shuttle, captured the world's attention; ignored are the thousands who die equally tragic and preventable, but duller deaths daily. Consider the following annual statistics:

- 43,500 killed in automobile accidents in the United States (1985)
- 1,384 murdered in New York City (1985)
- 36 murdered in Honolulu (1985)
- 150 killed in accidents in their own bathtubs (1984)
- 1,063 killed in boating accidents (1984)
- 3,100 dead from choking on food (1984)

—Newsweek, June 2, 1986

A single murder, that of hostage Leon Klinghoffer in 1985, commanded the front pages of almost every newspaper in the West for some days. Extreme political demands have been met because of the importance individuals, the media, and governments give to such tragedies. The threat of terrorism made millions of Americans change their travel plans in the summer of 1986. During the time it takes you to read this book, more people will die in automobile accidents in the United States than have *ever* (until the time of this writing) been killed by terrorists.

The way in which the mind caricatures reality to focus on the new and unusual is what makes terrorism good shock value. People will believe terrorism is a recurring strategy of the otherwise powerless, until they also realize its effect depends on the emphasis automatically given to it by the default positions of the "old minds" and, by extension, in the media. Terrorism taps into the nervous-system program that originated to register short-term changes in a steady state. A constant flow of murders is "tuned out," just as the sound of an air conditioner is tuned out shortly after it starts. But terrorism, or the discovery of a serial killer, acts like an occasional squeak in the air conditioner's motor; each time it sounds, it rivets our attention on the machine.

Nevertheless, biological evolution did its job well. It adapted our ancestors to their environments and, in that process, shaped us into rather special creatures.

Then how do we present the danger of greenhouse warming to the human being who must get the message about what we are doing to the Earth? It will involve, first of all, changes in how we write and how our media communicate. Surely, we always will need sensational stories to captivate us. But responsible newspersons can select equally exciting coverage of the major changes occurring in our world. Last summer, *Time* ran a story on 500 or so murders committed one week, who the victims and perpetrators were, what happened to them. Surely, when people are suffering from the effects of environmental pollution we can also run stories on what happens to these individuals. This is the key, since the mind responds only to representative examples. Shattered Romanian villages are prototypes of what might become of other places on the Earth, and we should see what happened in Romania. The democratization of Eastern Europe is of vast importance to us all. Yet a victim of acid rain or skin cancer is no less an object of sympathy than a victim shot dead in the street in a drug-related drive-by shooting. We need to dramatize and attend more to the continuing tragedies of our time than to the isolated "whale down an ice floe" story. But we need to know that the mind will always want to hear about the single hostage killed, the whales trapped in Alaska, Baby Jessica. We shouldn't fight that but should learn to put the message in the way our audience—the human animal—responds. An alliance of psychological and environmental scientists with the media may pay off far more than further brute-force communications. We must learn to acknowledge a changing world, before the only message possible is that we have become its willing victim.

ROALD Z. SAGDEEV
AND
SUSAN EISENHOWER

The Global Warming Problem: A Vital Concern for Global Security

Roald Zinnurovich Sagdeev is head of the theory division of the Space Research Institute of the USSR Academy of Sciences and a distinguished professor of physics at the University of Maryland. Academician Sagdeev worked on controlled fusion at the Kurchatov Institute of Atomic Energy until 1961, when he became head of the laboratory at the Institute of Nuclear Physics in Academgorodok, in Novosibirsk, Siberia.

Sagdeev was director of the Space Research Institute in Moscow from 1973 to 1988; during this time, the Institute coordinated many international space projects, including the 1986 Vega mission investigating Halley's Comet.

Academician Sagdeev is a people's deputy of the USSR. He recently wed Susan Eisenhower, granddaughter of the thirty-fourth president of the United States, Dwight D. Eisenhower.

Susan Eisenhower has worked for several years to establish groundbreaking programs for the promotion of U.S./Soviet exchange activities, policy discussions, and joint trade ventures. Her company, The Eisenhower Group, Inc., has been representing a number of major companies and organizations in the USSR since 1986. In addition, she played the leadership role in developing U.S.-Soviet policy conferences and forums at the Eisenhower Institute during the critical early years of *perestroika.* Through that organization, she and her husband recently established the Center for East-West Study and Initiative, to assess and encourage innovative approaches to East-West relations.

This spring, students of the University of Maryland went to the ballot box. Exhausted from the endless debate surrounding global warming, the students mischievously decided to put it to the vote: "Is global warming a myth?" The result of the poll, they surmised, would at least be conclusive.

While experts argue over the credibility of global warming scenarios and fine-tuned disagreements rage about how much the average temperature would increase during the next fifty years, there is no conflict of opinion that the science of global warming has to be strengthened and internationalized. Yet the variation of opinion regarding the level of commitment this phenomenon should receive is one of the great paradoxes of the modern era.

For the last forty years the United States and the Soviet Union have spent trillions of dollars to defend against the "worst-case scenario" of their national security. Military men who voiced gloom and doom were regarded as "cautious," and their skepticism was rewarded with substantial contracts to develop and deploy weapons systems capable of responding to even the most absurd eventuality. Ironically, today, those environmentalists who speak with alarm and deep pessimism about our preparedness to wage a war of survival against the threat to our environment and livelihood, are not labeled "patriotic" or "guardians of our security." They are frequently viewed as anti-growth, anti-*perestroika,* or worse: intellectually panicked.

Yuri Kariakin, a literary scholar and one of the Soviet Union's most outstanding contemporary philosophers, said recently, "It is far more likely that we will suffocate together before we . . . kill each other in a nuclear exchange." Such logic leads to the inevitable conclusion that among the imperatives of the post-postwar period is the need to elevate *ecological* survival to the level once the exclusive domain of *nuclear* survival.

Global warming has a crucial bearing on global security. Local or regional environmental hazards that have preoccupied almost every country during the last several decades are already undergoing an unavoidable process of de facto internationalization. The transport of Ohio Valley pollution across state lines in the form of acid rain, the poisoned Danube River, which crosses national borders; radioactive smoke from Chernobyl; marine pollution; depletion of the ozone layer—all these represent the internationalization of environmental problems and a violation of our common ecological home.

What stands in the way of an international consensus on an international cleanup? Perhaps it is the collective subconscious that dreads the day when the formidable price will have to be paid, while knowing that both sacrifice and political courage will be needed. When such an appraisal comes, it will also involve the messy international business of attributing blame and assessing financial responsibility.

Present-day problems aside, expenditures will also be necessary in the future, for capital investment that will develop new environmentally safe technologies and implement them worldwide. Without considering the massive scale required, even a semicomprehensive effort might be too little, too late.

RICH VERSUS POOR

While the United States constitutes less than 5 percent of the world's population, its share of carbon dioxide release is almost 25 percent. In strategic military confrontation, the principle accent has been on West versus East; in considering global ecological security, the major political and economic issue will be bipolar: rich versus poor.

The developed world must recognize the sobering reality that while it finds ecological issues formidable for itself, the underdeveloped world, particularly in the South, is far from being able even to contemplate mounting environmental and ecological dangers. Enormous shipments of military supplies directed from north to south must be replaced with environmental and ecological aid. This would require the creation of dedicated international funds, the transfer of special technologies, and the establishment of educational programs to develop human potential. Without ever-growing support, the developing world will not be able to stay afloat on its own against the relentless escalation of ecological degradation. Eventually, the developed world could find itself as an ecological hostage.

The trend has already begun.

Sweden, one of the most environmentally aware countries in the world, has been grappling with cross-boundary pollution from Poland. It is now exploring how much it should invest in Poland to clean up industrial sectors there that are exporting dangerous fumes from antiquated smokestacks—in the form of acid rain—to the thick forests of Sweden.

In 1989, Andrei Sakharov and Elena Bonner proposed the establishment of an international trust fund that would raise money to purchase the Amazon rain forest as a protected reserve. Sakharov and Bonner were concerned about the Brazilian government's slash-and-burn policy, and many prominent scientists, bankers, policymakers, and businessmen are likewise looking at ways to pay a form of unavoidable ransom for a reasonable environmental policy to save the region and the "Earth's lungs."

ENERGY EFFICIENCY OF VITAL IMPORTANCE

Energy efficiency has become a critical issue, not only because of the resultant fuel savings but also because of the impact it will have on pollutants that enter our atmosphere.

While the developed world increases its energy production (and its carbon dioxide releases grow at the rate of 1 percent annually), the Third World consumes 5 percent more energy per year (with a 5-percent increase in carbon dioxide releases). Struggling for economic survival, the Third World often uses the cheapest methods to produce and use energy. This creates a double necessity to develop and implement energy-efficient technologies now.

Sixteen years have passed since the first oil crisis that forced Japan and the West to invest in energy conservation and efficiency. And that has brought dividends. Maybe most important of all, the last decade proved that national economies in developed countries could grow without an accompanying increase of energy production. At the same time, the Soviet Union, with its rich national resources of oil, gas, and coal, was not motivated to implement a similar plan. Now it burns, annually, approximately 2,000 million tons of counting units, called "conditional fuel" (counted in energy efficiency of coal). If this grandiose energy infrastructure would have the same efficiency as that of systems in Western countries, according to Soviet analysts, the annual savings could be as high as 700 million tons, bringing the total down to 1,300 million tons.

The savings would be a boon for the economy, and would represent a significant deceleration of carbon dioxide releases into the atmosphere.

Therefore, the long-term imperative—and challenge—is to develop and implement special environmentally sound technologies, which would minimize hostile wastes and pollution. For global warming, the key to a solution lies in increased energy efficiency and conservation for fossil-fuel cycles, and eventual use of alternative energy sources. Nuclear power at present is under strong attack because of two key unresolved questions: what to do with radioactive wastes; and how to make nuclear reactors inherently safe. Without international efforts to bring back nuclear power as a reliable source of energy (which must be supported by the International Atomic Energy Agency), modern civilization would have to survive on fossil fuels and the current state of atomic energy.

URGENT ACTIONS

In the meantime, other urgent actions are required. Stronger research should be conducted, accompanied by increased activity on a grass-roots political level. And if we would recognize the need for it, dedicated ecological summits should also take place within both bilateral and multilateral frameworks.

In the expected wave of deep cuts in military budgets, and widespread conversion from military to civilian industries, special plans to look at environmental technologies as a new target of activity are necessary. In addition, significant work must be undertaken to establish an international framework not only to reach treaty agreements on emissions and standards, but also to mobilize international science to pioneer developments on a bilateral or multilateral basis.

By any standards, logical steps could be considered—at least on the level of so-called minimal deterrence. (In the terminology of military strategists, this is the minimum level of the nuclear arsenal capable of deterring a potential aggressor.) In the case of the environment, the barest necessity, again, should be research and monitoring to enhance the reliability of risk assessments and a recognition of our danger. A threat as great as global warming does not present a win-or-lose situation. We must keep temperature increase as low as possible. It could be done through analysis of comprehensive global scenarios for actions to minimize the damage. Finally, one of the most important objectives: to make political decisions on national and international levels. To do this, expanded efforts at public education and political awareness will be required.

A LONG-TERM ISSUE OF SURVIVAL

At times it is difficult to feel optimistic about mankind's ability to mobilize for the fight to save our environment. Economically, for instance, we have demonstrated a preference for short-term expediency, rather than the long-term fiscal responsibility that might have saved our children and grandchildren from the nightmares of crushing indebtedness.

Environmental destruction is the same kind of long-term, slowly evolving problem. Like economic bankruptcy, global climate change may sneak up on us if we are not vigilant and if we do not take action now. But unlike economic irresponsibility, which can be corrected (however difficult that may be), irresponsible treatment of the environment cuts to the heart of life itself. We have to recognize the imminent paradigm for all the future generations on the earth. Not only are terrestrial resources finite, but they are also infinitely fragile.

The physicist Leo Szilard, in his sad science fiction satire, "Report on Grand Central Terminal," painted a story of extraterrestrials attempting to discover why this civilization destroyed itself in a nuclear holocaust. What kind of conflict eventually triggered the global war? While the majority of the expedition thought it was a conflict among different races, one young economist, Xram (Marx) found, in one section of Grand Central Terminal, metal disks in what had apparently been public toilets. He theorized that a society based on the principle that it was necessary to earn disks (coins) to deposit its wastes had to be unstable—and likely to come to a tragic end.

Let us hope understanding prevails before we engage in our own conflict based on the idea that only those countries with money can effectively dispose of their environmental waste. Science fiction perhaps, but not completely impossible when one considers the problems we are dealing with today in the disposal of solid and nuclear wastes.

We must come to the recognition that the Earth has a finite carrying capacity. It cannot limitlessly support its growing population, and in the final turn there is no way to avoid the most outstanding issue of all: global family planning. After all, our little planet is a lonely spaceship Earth, moving in the vast galactic spaces. As any kind of spaceship launched to carry a certain payload and crew, it cannot permanently accomodate an ever faster-growing cargo. The progress of science and technology can do wonders, but, sad as it may sound, ecological catastrophe is unpreventable in the next century without a conscious approach to the limitation of our own species's growth.

"Interdependence," a key word for internationalists during the last several decades, found form in such brilliant concepts as "nuclear winter"—only rhetoric and theory until the Chernobyl nuclear disaster registered radioactivity around the globe.

Even with the inescapable reality of interdependence, people the world over will have to solve global problems despite those who view nationalism as the most important consideration. Opponents of strong action for global warming preparedness for instance, come up from time to time with a most peculiar form of optimistic scenario. Said one leading Soviet expert at a conference we recently attended, "Why should we worry about global warming at all? Siberia would be turned into paradise, and that would be good for the Soviet Union." To that, another Soviet colleague standing nearby quipped ironically: "Great, Kansas would become a desert. *Then* where would we get our grain?"

OREN R. LYONS

A Respect for the Natural Law

Oren R. Lyons is the faithkeeper of the Turtle Clan of the Onondaga Iroquois nation, and associate professor of Native American Studies at the State University of New York at Buffalo. Lyons has represented the Iroquois nation, and native peoples in general, at many local, national, and international conferences and symposiums on issues ranging from aboriginal rights to Native American art. Lyons has also represented the Onondaga Nation at Native American councils and treaty meetings on national and international levels. This year, he was a member of the planning and steering committee for the Global Forum of Spiritual Leaders for Human Survival, held in Moscow. During this conference, Lyons met with Soviet President Mikhail Gorbachev to discuss major environmental issues. He is an accomplished lecturer, athlete, and artist, and has written books and articles on the rights of indigenous peoples, Native American law, and Native American history.

*You must teach your children that the ground beneath their feet is the ashes
of our grandfathers so that they will respect the land. Tell your children that
the earth is rich with the lives of our kin. Teach your children what we have
taught our children, that the earth is our mother. Whatever befalls the earth
befalls the sons of the earth. Man did not weave this web of life. He is merely
a strand of it. Whatever he does to the web, he does to himself.*[1]

—CHIEF SEATTLE, 1854

All my relations—for we are indeed related—we share a common destiny for good or ill. Our fate is intertwined with the life on this planet we call Etenoha, our mother.

The Greenhouse/Glasnost Symposium, held in Sundance, Utah, in August 1989, brought the enormity of the environmental problems facing the nations of the world into sharp focus. Less focused were the more disturbing conclusions about what to do about those problems. The brilliance, energy, and commitment of the symposium participants were impressive, but they only underscored that more work must be done to educate world populations in united efforts of conservation and sacrifice if life as we know it is going to survive.

A letter of understanding between the USSR's Academy of Science and the Institute for Resource Management in the United States outlined bilateral action needed to begin to address issues related to global warming. Further letters to President Mikhail Gorbachev and President George Bush recommended four specific actions to engage the Soviet Union and the United States in the issues and problems of survival on this planet. As one would expect from the scientific community, the proposed actions were sharp and clear (see appendix). All of this achieved as a result of the symposium.

As American Indian delegates, we were concerned with what was not addressed so clearly and what we feel to be necessary for survival. I will restate it here in this impressive collection of essays by scientists, policy-

---

1. Hans Koning, *Columbus: His Enterprise* (New York: Monthly Review Press, 1976), p. 1.

makers, and businessmen: the understanding of the *spiritual laws* that prevail over all life.

We believe the genesis of the crisis the world faces today is a moral one, that the problems of the environment are related directly to the conduct of human beings as they despoil and ravage the resources of the Earth in the name of economic development and progress of industrial states. Indeed, it was agreed that the USSR and the United States are responsible for 45 percent of pollutants contributing to global warming. My relations, listen to what we American Indians have to say. We believe the battle for survival is waged in the minds of men, and it is what is in the minds of men that has brought us to this particular battlefield.

In the Americas the conflict began with the landfall of Christopher Columbus, on his quest for "glory, God, and gold." Columbus enslaved hundreds of Indians, and in his thirst for gold he had the hands cut off Indians who failed to bring in their three months' quota of it. In forty years' time the Spanish wiped out an entire race of people from the island of Hispaniola. Thus began the Indians' "education" about the avarice of the white man.

The pain endured by the native peoples of the Americas at the hands of the colonizers needs to be acknowledged by their descendants so we can all, white, black, and red, begin the healing process and address the fundamental evil—greed.

The "discovery" of America is in itself a theft of the indigenous peoples' land and heritage. The word *discovery* implies uninhabited lands—virgin territories, unoccupied. In truth, these lands were and still are inhabited by their ancient aboriginal peoples and nations.

This attitude of discovery and, therefore, ownership of Indian lands extended to all white men, Spanish, French, English, and others. The Puritan captain Endicott, in an expedition from Salem against Thomas Morton's Maypole dancing, was advised by his chaplain, Elder Palfrey, not to consider Indians harmless:

> There are three thousand miles of wilderness behind these Indians, enough solid land to drown the sea from here to England. We must free *our* land of strangers, even if each mile is a marsh of blood.[2] *(Emphasis mine.)*

The history of the Americas is one of blood and violence, of conquest and slavery. It must be told to children today to bring balance, order, and true peace to our future generations.

2. Richard Drinnon, *Facing West: The Metaphysics of Indian Hating and Empire Building* (Minneapolis: University of Minnesota Press, 1980), p. 4.

Our nation, the United States of America, cannot meet a moral crisis without challenging every aspect of it. We American Indians believe the motivation of greed and power that commercial enterprise brings continue to be at the heart of the national dilemma and our response as a nation to the effects of global climate change.

Perhaps the words and wisdom of the American Indian grandfathers can still offer another direction to explore.

Is it possible for a linear civilization to understand the layers upon layers of ritual custom and songs of the ancient ones performed to fathom the rhythms of these lands? A vital function of ceremony includes thanksgiving, respect for all of life and its processes, the need for balance, and the involvement of community in celebration. The Jesuits complained:

> The greatest opposition that we meet . . . consists in the fact that their remedies for diseases, their greatest amusements when in good health; their fishing, their hunting, their trading; the successes of their crops, of their wars, of their councils—almost all abound in diabolical ceremonies.[3]

They did not understand the role of ceremony in American Indian societies. People today continue to be mystified by American Indian ceremonies; indeed, ceremonies and customs of indigenous peoples around the world have been a source of curiosity to the so-called civilized world for as long as they have been observed and interpreted.

For our people, ceremony is community. It is a way for us to celebrate, in thanksgiving, the gifts of the universe—those life-giving forces of the natural world that interact to bring us the dawn, the soft southern breeze of spring, the thundering voices of our grandfathers who bring the rain.

Ceremonies for the endless cycles of life have brought us to this time in the history of human beings on the Earth. We are told by the old people that we have failed before, that this, our life, is a story without end if we obey the natural laws that prevail over all creation, if we give thanks individually, and in community.

Our children learn by example, they watch the actions of their families intently with intelligence and emulate them as a symbol of love and respect. If Grandmother and Grandfather attend ceremonies, if Mother and Father, aunts and uncles, cousins and friends are there in celebration with the chiefs and clan mothers who lead the thanksgiving, then it must be good

---

3. Olive Patricia Dickason, *The Myth of the Savage and the Beginnings of the French Colonization in the Americas* (Edmonton: University of Alberta Press, 1984), pp. 251–252.

and right. *I too shall carry my children to these great spiritual events,* our children learn. And we pass on to coming generations the ceremonies that teach respect for all life-giving forces, and join our relations in spiritual song of endless celebration.

My relations, I shall continue. Our brothers from across the sea did not understand or believe in our natural laws. They despoiled our towns, scattered our people, and hunted down our leaders, and killed or enslaved them. John Elliot of Plymouth Colony wrote:

The design of Christ is not to extirpate nations, but to gospelize them. To sell souls for money seemeth to me a dangerous merchandise. . . . There is a dark cloud upon the work of the gospel among the poor Indians.[4]

My relations, I shall continue: In spite of this long, dark history we have survived, and we say to you that perhaps there is a reason. In these troubled times, as men and women gather all over the Earth in councils to discuss the dangers surrounding us and assess the damage wreaked collectively on our Mother Earth, as fear grows in the eyes of our children, we look to our*selves* and see that we have all been hypocrites, we have lied to ourselves and made laws condoning actions that strip the Earth of its natural resources.

Government has made laws that allow this generation to profit at the expense of its grandchildren. We say that creates a *moral crisis.* Greed has been cleverly disguised as law. Our national leaders have lied to the American people.

Science can point out dangers, but science cannot turn the direction of minds and hearts. That is the province of spiritual powers within and without our very beings—powers that are the mysteries of life itself.

Perhaps now is the time to listen to the voice of the Indian, to look to the original peoples of these lands, this great turtle island.

The words I speak may be the collective position of the people of the Onondaga Nation. They may at times be the voice of the grand council of chiefs of the Haudenosaunee, or they may be the wisdom of the traditional circle of elders and youth in North America—good minds.

My relations, hear our voices.

Our grandfathers spoke of the crystal-clear waters of springs, streams,

4. Angie Debo, *A History of the Indians of the United States* (Norman: University of Oklahoma Press, 1970), p. 49.

rivers, lakes, and great inland seas. They spoke of the fresh, pure waters. The first law of life is water.

They spoke of ancient trees, grandfathers of another age, trees so huge it took six men to circle their trunks.

They spoke of a vast forest with leaves so thick that sunlight barely found its way to the forest floor: a squirrel could travel from the great eastern sea to the Mississippi River without touching the ground.

They spoke of flowers and medicines growing in profusion beside the fruits, nuts, and berries, feeding and nurturing not only human families but also the animal nations that abounded and prospered in these vast lands.

They spoke of fish so abundant that in spawning season the streams and rivers were so full a man could run on the fishes' backs.

They spoke of passenger pigeons so plentiful their roosting places were stripped of limbs, their combined weight breaking those limbs. So plentiful, they darkened the sky for hours when they migrated.

They spoke of vast herds of game, deer, elk, and massive buffalo roaming the entire continent, powerful and endless. But they did end, and so our grandfathers received another lesson about the avarice of the white man.

My relations, listen as we continue.

The lesson our people learned was that man wanted to dominate, and what he couldn't dominate, he destroyed. Mankind was capable of destroying life, his own and that of the natural world. Our people—so closely aligned and intertwined with the order of the natural world—suffered the same fate as the trees and the wolves, our spiritual relatives.

The lesson taught us that mankind could be motivated to exploit natural resources, and the environments these resources provide, to the point of total depletion and extinction of animal and fish life.

The lesson illustrated to us that there were people who were ignorant of our natural law, *the* natural law, or who chose deliberately to ignore it. Our people gathered together in alarm and held the principles of the great natural law to their bosoms. They fought to protect the ceremonies that celebrated these principles and ensured the existence of the generations to come.

My relations:
The natural law as we understand it is the ultimate authority upon these lands and waters. It is the prevailing law of life and the order of life upon this Earth we call our mother.

It is the law the Creator put here, set down deliberately, firmly and with finality, to govern all life in this creation.

This is how we understand it: The Great CREATOR planted life on the Earth. He planted all nations of life, from the grasses to the trees, from the insects to the elephants, from the tiniest life in the waters to the great whales.

And he planted the families of mankind, in the four great sacred colors of black, white, red, and yellow.

He gave instructions to these great nations of life, from the grasses to the whales, and they continue to follow them to this very moment. They carry out their duties as best they can. They live in a state of grace. They do no wrong.

To human beings, the Great Creator gave additional responsibilities. He gave us hands to work with; he gave us intellect and the power of reason; he gave us options to choose our paths—to do what is right or what is wrong. He gave us foreknowledge of death and insight into life after death. And the Creator gave each of us a mission in this life that is ours alone. These are responsibilities, not gifts, to be cherished and shared for the benefit of *all* life.

My relations, this is what we believe. We share it with you so you may understand us better. These are our cosmologies. Your stories may be different, but we believe that we all received the same instructions at the creation of life.

The natural law is a spiritual law. Its powers are both light and dark. We are blessed and we prosper if we live by the law. Existence is dark, terrible and merciless, if we transgress it. There is no discussion with the law, there is only understanding and compliance. Its tenets are simple:

A respect for all life, for all life is equal.

Thanksgiving ceremonies for the special forces of nature:

A thanksgiving for the thundering grandfathers who water the earth, and the people who freshen the springs, streams, lakes, and rivers.

A thanksgiving for the four winds who bring the seasons and sow the seeds of life.

A thanksgiving for the corn, beans, and squash that sustain our lives and give strength to our bodies.

A thanksgiving to our grandmother the moon, who raises and lowers the tides of the great salt seas, who gives us light at night, who marks the cycles of female life and the seasons.

A thanksgiving for our elder brother the sun, who unites with our mother the Earth to bring forth life in all seasons, and who brings us light each day as we wait in the morning to greet him.

A thanksgiving to the stars in their infinite wisdom, who give us direction at night (most of which we have now forgotten). We give thanks for their beauty, and the dew they bring in the night.

A thanksgiving to spiritual beings assigned to help human beings carry out their duties.

A thanksgiving to the Great Creator, the master of all life, for the creation of the universe, and the Earth we have been given to enjoy and protect, so that seven generations from this day our children will enjoy the same resources we enjoy now.

Listen to the howl of our spiritual brother the wolf, for how it goes with him, so it goes for the natural world.

My relations, so now we will continue.

We gather from the four directions of the Earth to report on how it is where we come from. This news is heavy. There seems to be a determined effort to destroy life on this planet.

My relations, my friends and colleagues, we are here at this time in history with a task we cannot leave to our children. Our choice requires courage, fortitude, and a will inspired by understanding the great spiritual law of our Mother Earth. Heed the word of our grandfathers, who instructed us:

"Take care how you place your moccasins upon the earth, step with care, for the faces of the future generations are looking up from earth waiting their turn for life." So the decision is simple. We must obey the natural law, or perish.

TERRELL J. MINGER

Epilogue:
The Unresolved Question
of the Decade

Terrell J. Minger is president and chief executive officer of the Institute for Resource Management. Before joining IRM in 1987, Minger served as deputy chief of staff for the State of Colorado. He has also served as assistant city manager of Boulder, Colorado, and as city manager of Vail, Colorado. Minger was a Loeb Fellow in Advanced Environmental Studies at Harvard University. He has edited two books, *Growth Alternatives for the Rocky Mountain West* and *The Future of Human Settlements in the West*.

G reenhouse/Glasnost was originally conceived by Roald Sagdeev, former director of the Space Research Institute of the USSR Academy of Sciences, and Dr. Walter Orr Roberts, founder of the National Center for Atmospheric Research. It took its original form early in 1987 as the first permanent computer teleconference between U.S. and Soviet scientists and environmental-policy experts. The most advanced technology of the information age was put to its highest possible use—the free, bilateral, uncensored exchange of knowledge about global warming. A threat to the future of the planet that some say rivals nuclear holocaust, global warming is truly one of the unresolved issues of the decade. Greenhouse Glasnost is as much a symbol of our rapidly changing times as a label defining a common problem and opportunity.

Greenhouse/Glasnost, the symposium sponsored by the Institute for Resource Management and the USSR Academy of Sciences and held in August 1989 at Sundance, Utah, was an outgrowth of this early electronic exchange. Its timing was fortuitously in sync with major paradigm shifts in both countries—the "new thinking" going on in the Soviet Union as well as the spreading environmental awareness on both popular and corporate levels in the United States.

As early as the 1970s, Soviet environmentalists were able to organize protests—condoned by the state yet nonetheless very effective—against damage to the natural environment. The most prominent among these protests involved an effort to save Lake Baikal from further industrial pollution. Today the fledgling Soviet environmental movement continues to be one of the largest and most effective voices of *glasnost* and *perestroika*.

In fact, during the last decade or so, concern about environmental protection in the USSR has taken on a much higher priority. This is a reaction to years of neglect and a single-minded state focus on *uskorenie,* or national accelerated development and industrialization. The Soviets refer to the environment fondly as *priroda,* literally "nature," a term that covers all aspects of life within the biosphere. Every visitor to the Soviet Union comes away deeply impressed with the people's love for *Rodina,* the Motherland. For what is patriotism but the love of one's country, the love of one's land? It is interesting to note that as early as 1968, the physicist Andrei Sakharov,

Nobel laureate and Soviet conscience, warned that "the salvation of our environment requires that we overcome our divisions and the pressure of temporary local interests. Otherwise the Soviet Union will poison the United States with its wastes and vice versa."[1] Sakharov pleaded for U.S.–Soviet cooperation and the creation of a "Law of Geohygiene" to deal with environmental problems on a regional and global basis.

Despite almost twenty years of U.S.–Soviet technical and scientific cooperation on environmental problems, the public of either country is largely ignorant of the views or efforts of the other to protect the respective homeland ecology. Recent collaboration continues a little-known but long-standing tradition, however. Mikhail Lomonosov, the founder of Russia's Academy of Sciences, eagerly exchanged information with American colonial scientists Benjamin Franklin and Ezra Stiles in the eighteenth century. Ever since, with some interruptions, there has been a bilateral exchange of ideas on science, space, and the environment. But only recently, with the thaw in the Cold War, have political leaders of both countries begun to recognize that environmental problems are not unique to East or West, and are not the exclusive by-products of communism or capitalism. On the contrary, environmental problems increasingly require global cooperation and often multinational, transboundary solutions.

President Mikhail Gorbachev illustrated this change in world view at the January 1990 Global Forum held in Moscow. Paraphrasing Immanuel Kant, he stated that the "ecological imperative" has entered the political reality of all nations and people. He admitted the Soviet Union had only recently come to appreciate the importance of ecology at a high-policy level and attributed the country's ecological carelessness to its size and its vast riches. Gorbachev offered a simple, yet revolutionary plan for the global environment, which included six major points: (1) creation of an international code of environmental ethics; (2) preservation of unique ecological zones that have planetary significance; (3) development of international ecologically clean technologies; (4) development of an international mechanism to monitor the biosphere; (5) creation of an international "Green Cross" to deal with global ecological disaster and administer ecological first aid; (6) renewal of the proposal for the Soviet Union to cease all nuclear testing when the United States is ready to do likewise.

Greenhouse Glasnost is a powerful as well as a predictive metaphor. It reflects the underlying belief shared by scientists of both nations that each of us needs to approach an issue as complex and uncertain as climate change

---

1. Andrei Sakharov, *Progress, Coexistence and Intellectual Freedom* 49 (1968).

with a cooperative, open, yet skeptical mind.

Currently, environmentalists and businessmen argue the costs and benefits of international action on global warming while scientists and politicians debate whether politics and public opinion have outpaced scientific understanding. Ongoing debates and uncertainties notwithstanding, the issue of whether the greenhouse effect is bad, beneficial, or benign is beside the point.

Even in the absence of absolute scientific consensus, there is sufficient evidence to take the threat of global warming very seriously. The first concern and present danger discussed over and over in this series of essays is the potential for fiddling while Rome burns—in the present case, while the planet heats up—and the climate is irreversibly altered.

The second concern is the rapidity and scope of change that is being forecast—a rate unprecedented in the history of mankind. Throughout these discussions, the message that comes through loud and clear, whatever one's perspective, is the rare and perishable opportunity for mankind to be proactive. We have the collective capacity—not one institution or nation alone, but through the efforts, research, and findings of many institutions and nations—to increase our understanding of the climate problem. Someday soon we hope to be in a position to predict its implications and then to apply reason and forethought to its resolution. Global climate change is only a partly decoded message that enables all of us in the United States and the Soviet Union to think about and prepare for our common future: not to wait until all the science is in full agreement, but to face the issue of global change head-on, together, in a problem-solving frame of mind.

Of course, many scientific, economic, and policy questions must be answered. We need to be able to predict climate impacts on very specific geographic locations. We need to know the costs and benefits of various strategies for arresting or slowing the buildup of greenhouse gases. We need to discuss appropriate regulatory standards and cost-effective measures. But we need to pursue these tactics simultaneously, not serially. Even as scientists are refining their models, databases, and predictive capabilities, politicians should be examining coping strategies; businesses should be factoring state-of-the-art knowledge into their corporate strategic plans; and all of us, as individuals, should be changing our habits as consumers and voters. In many respects, people in both the United States and the Soviet Union are more willing to deal with changes in life-style than their governments are. We need to position ourselves, because eventually we will have to act. How, when, and with what force of commitment—these constitute the unresolved question of the decade.

Climate change will occur regardless of our ecological and economic preparedness. We can wait for the normally lethargic process by which science makes its way into public awareness, and eventually alters the way we go about our political and economic business. And we will likely find ourselves in a state of perpetual crisis management. Or we can muster our collective will and speed up the climate-saving process. We can adapt consciously rather than react instinctively.

According to the Worldwatch Institute's *1990 State of the World,* the planet has about forty years to come to terms with the necessity for an environmentally sustainable economy or it will plunge into a headlong economic and environmental decline. The report offers elements of a pre-ferred economy that could be sustained over many decades. In general, these recommendations corroborate the suggestions of the contributing authors of *Greenhouse Glasnost.* They include reducing world population growth below projected levels, stabilizing emissions of industrial and green-house gases, and eliminating the loss of forest and plant cover and the reduction in the diversity of animal species.

If a sound economy is achieved over the next forty years, it will rely heavily on recycled materials and much less on virgin materials. The deforestation of tropical rain forests will have ceased and much of the world will be reforested. While these conditions appear to be major departures from the status quo, the Worldwatch Institute report states clearly that most changes are possible without major shifts in the social, economic, and moral character of any nation.

In the face of the recent unprecedented cascade of international events, it is hard to deny that fundamental changes and restructuring of the world order as we have known it for the last fifty years are at hand.

Peace seems to be breaking out everywhere, democracy and capitalism are flowering in heretofore unlikely places, and the superpowers have an opportunity as well as a global responsibility to redirect their competitive energies to the problem of fostering joint economic *(uskorenie)* and environ-mental *(priroda)* as well as military security.

Paul Kennedy, in *The Rise and Fall of the Great Powers,* ascribes the decline of great nations of the last century such as France, Spain, and Great Britain to economic and military overextension and taxation. *Now* is a perfect time to examine and debate the larger questions of what is possible economically and ecologically when the two giants turn their collective minds and resources to global peace and security.

The fundamental question of how a modern industrial society, in a time of relative peace, chooses whether to spend a significant proportion of its

Gross National Product on defense, or on real social, economic, and environmental problem-solving is a matter of political priorities and practical politics. *Greenhouse Glasnost* offers constructive insights to the dilemma of which course the United States and the Soviet Union will take: Will they address issues related to global climate change, agriculture, and economy, or will they continue to pursue plans such as Star Wars and further arms buildup?

Indeed, the roller coaster of world events seems propelled by an uncertainty—a shaky but correct moral compass that points toward facing global ills common to all rather than perpetuating longstanding national prejudices!

The bottom line is that scientists cannot *ever* give us the absolute definitive answers politicians and businessmen want in considering whether or not to take action on global climate change. But there seems to be an important overreaching and widely held conclusion that most of the measures we would begin take to mitigate the possible impacts of climate change—we should be taking now—willingly, and for rational reasons. We need energy conservation—greenhouse effect or not. We must cut back on our use of fossil fuels, not to freeze in the dark but to curb air pollution generally. We need to preserve our forests and animal habitats.

Developed countries must help developing countries advance economically in ways that preserve our planetary home. Strong leadership is required for the United States and the Soviet Union to compete peacefully as we determine what to do about forecasts of global change. Since our world now is altered more in a decade than it once was in a millennium, the most important concept is the understanding that the only constant is change itself.

An old Russian saying has it: "I need not fear my enemies because the most they can do is attack me. I need not fear my friends because the most they can do is betray me. But I have much to fear from people who are *indifferent.*"

Human indifference and passivity represent the greatest threats to planetary survival. The "ecological imperative" demands that we care and act.

Afterword:
Global Warming
A Parable

Once upon a time, there was a small planet that suffered from a problem of the gradual heating up of its atmosphere. As industrial processes, automobiles, and people dumped their wastes into the air, a blanket of pollution formed a greenhouse around the earth.

The two countries responsible for half of the problem were engaged in business as usual. Clearly something needed to be done, but both countries were busy building up their nuclear arsenals and putting out domestic political fires. They claimed that there weren't enough resources to study the problem fully, and that there was insufficient public support for devising strategies to reduce pollution or increase energy conservation.

By the year 1990, things went from bad to worse, so the presidents of both countries got together and decided more study was needed before any course of action could be embarked upon. They asked their advisors for advice. Their advisors said the answer was simple and politically very palatable. "What is needed is more information. Everything is connected to everything else. If global warming is to be solved, a series of international global warming conferences must be held."

After the conferences, scientists, policymakers, and politicians would study, computer-model, and cost-benefit the possible options further . . . after which the United Nations and the world community would be consulted . . . after which the problem would be restudied and re-computer-modeled and re-cost-benefited on the basis of each nation's contribution to the problem and the solution. . . . Everything addressed and readdressed.

Meanwhile, world population expanded, forests were cut back, and the concentration of carbon dioxide and other gases in the atmosphere continued to increase at alarming rates.

Unfortunately, by 2050, there was still no consensus about whether it was happening, never mind what should be done about it. Nations turned inward as each engaged in internal struggles with massive flooding of major population centers, unmanageable migrations of environmental refugees, drought, and famine.

Humans and many other forms of life on earth ceased to exist, and the planet breathed a sigh of relief as its temperature began to cool down.

Appendix:
An Open Letter
to Our Presidents

Our world is changing at a staggering rate. As old walls crumble, new alliances emerge and with them, new hope for our future.

As leaders of the two nations who, in their rivalry, have kept the world on edge, you have both the opportunity and the responsibility to help the world craft a new way of looking at the planet that will guide us safely into the next century.

We need a new concept of international security that recognizes the need for environmental stability.

The Malta meeting is an historic opportunity. As you focus in your informal discussions on recent political events in Eastern Europe and elsewhere, don't forget that our future will ultimately be determined not only by how we treat one another, but also how we treat the life support systems of the planet.

In August 1989, the Institute for Resource Management and the Soviet Academy of Sciences convened a diverse group of U.S. and Soviet leaders in Sundance, Utah, to discuss Global Climate Change. The following is a public reprint of a letter to both of you from many of the individuals who gathered in Sundance. We urge both of our nations to demonstrate leadership on this important issue.

August 26, 1989

Mikhail Gorbachev   George Bush
Chairman, Supreme Soviet   *President*

Recognizing the commitment you have pledged toward the environment, we wish to send you the conclusions from a symposium held at Sundance, Utah, convened by the Institute for Resource Management and the Soviet Academy of Sciences.

A symposium of American and Soviet scientists, policymakers, environmentalists, industry leaders and artists met to discuss bilateral action on perhaps our planet's gravest environmental problem: global climate change.

The improved relations between the two powers have created new opportunities for joint efforts to reduce the possibility of catastrophic global climate change. The USA and the USSR are the two largest producers of greenhouse gases. The USSR and the USA are also the two principal sources of the world's scientific knowledge which can be deployed to restrain global greenhouse emissions. Therefore, we should provide the leadership in the search for common solutions to the global warming problem, as well as for the environmental security of the world.

The Sundance Symposium arrived at four specific recommendations which we hope you will consider and support. We believe the struggle for the prevention of greenhouse gases should be raised to the level of international security. We propose the formation of an environmental security alliance between our two countries. We further propose our two nations work together on these four top priorities:

- To promote energy-efficient technologies which are both cost-effective and nonpolluting by enlisting the appropriate existing agencies in the two countries.
- To phase out the production and use of all CFCs as rapidly as possible, and no later than the year 2000, and to substantially reduce CO_2 and the remaining greenhouse gases.
- To reduce the rate of deforestation worldwide, and initiate reforestation efforts by enlisting the appropriate existing agencies in the two countries.
- To create citizen participation in the above efforts through a series of joint educational programs and by a direct appeal for action by each head of state.

We realize the success of these initiatives lies in engaging the efforts of all citizens. We stand ready to assist you in educating the public, as we were educated by the Native Americans who opened our conference:

*We do not inherit the land from our ancestors,
we borrow it from our children.* ”

Respectfully yours,

John Adams, Executive Director
National Resources Defense Council

Bill G. Aldridge, Executive Director
National Science Teachers Association

Howard P. Allen, Chairman of the Board
Southern California Edison Company

Governor Cecil Andrus
State of Idaho

Dr. Daniel B. Botkin, Professor of
Biology and Environmental Studies
University of California–Santa Barbara

Senator Bill Bradley
State of New Jersey

Dr. Noel Brown, Special Representative
of the Executive Director, United
Nations Environment Programme

Robert Bunting, Director of Corporate
Affiliates Program, University
Corporation for Atmospheric Research

F. Scott Bush, Visiting Fellow in
Environmental Affairs
Center for Strategic and
International Studies

Joseph A. Cannon, President and
Chief Executive Officer
Geneva Steel Company

Dr. Edward A. Dalton, President
and Chief Executive Officer
The National Energy Foundation

Dr. Devra Lee Davis,
Scholar in Residence
National Academy of Sciences

Mark Dayton, President
Vermilion Investment Company

Michael Deland, Chairman
President's Council on
Environmental Quality

Ann Druyan, Secretary
Federation of American Scientists

Dr. Paul R. Ehrlich, Bing Professor
of Population Studies
President, Center for Conservation
Biology, Stanford University

Susan Eisenhower, President
Eisehower Group, Inc.

Daniel J. Evans, Former U.S.
Senator and Governor
State of Washington

Michael Fosberg, Forest Fire and
Atmospheric Sciences Research
U.S. Forest Service

Dr. Peter Gleick
Pacific Institute for Studies in
Development, Environment and Security

Peter C. Goldmark, President
The Rockefeller Foundation

Gilbert Grosvenor,
President and Chairman
National Geographic Society

Dr. Jay D. Hair, President
National Wildlife Federation

Dr. James E. Hansen
Atmospheric Physicist

Dr. Alan D. Hecht Climatologist

Dr. Richard A. Herrett
Government Relations Scientific Liaison
ICI Americas

Phil Hogue, President
Daniels and Associates

Dr. William W. Kellogg,
Senior Scientist, Retired
National Center for Atmospheric
Research

Charles Kittrell,
Executive Vice President,
Retired
Phillips Petroleum Company

Frederic Krupp, Executive Director
Environmental Defense Fund

Justin Lancaster, Executive Director
Environmental Science and Policy
Institute

Dr. Daniel A. Lashof, Senior Scientist
Natural Resources Defense Council

Dr. Stephen Leatherman, Director
Center for Global Change

Dr. James M. Lents, Executive Officer
South Coast Air Quality Management
District

Dr. Thomas E. Lovejoy, Assistant
Secretary for External Affairs
Smithsonian Institute

Amory Lovins, Director of Research
Rocky Mountain Institute

Oren R. Lyons, Faithkeeper
Onondaga Nation

Tom Mathews, Partner
Craver, Mathews, Smith and Co., Inc.

Terrell J. Minger, President
Institute for Resource Management

Dr. Irving Mintzer, Director
Center for Global Change

George Montgomery, President
Montgomery Properties, Inc.

Richard D. Morgenstern, Director,
Office of Policy Analysis
Environmental Protection Agency

Dr. Michael Oppenheimer, Senior Scientist
Environmental Defense Fund

Dr. Robert Ornstein, President
Institute for the Study of Human
Knowledge

Dr. Larry Papay, Senior Vice President
Southern California Edison Company

Stephen Pezda, Principal Research
Engineer Associate
Ford Motor Company

Barbara Pyle, Environmental Editor
Turner Broadcasting System Inc./CNN

Robert Redford, Founder
Institute for Resource Management

Dr. Walter Orr Roberts, President Emeritus
University Corporation for Atmospheric
Research

Brian A. Rosborough, President
EARTHWATCH

Dr. Carl Sagan, Astronomer
Cornell University

Academician Roald Sagdeev, Member
Academy of Sciences of the USSR
Congress of the People's Deputies
of the USSR

Congresswoman Claudine Schneider
State of Rhode Island

Dr. Stephen Schneider, Head of the
Interdisciplinary Climate Systems Section
National Center for Atmospheric
Research

Russell L. Schweickart, President
Association of Space Explorers

Jeff Sirmon, Deputy Chief for Programs
and Legislation
U.S. Forest Service

Payson R. Stevens, President
InterNetwork, Inc.

Maurice Strong, President
World Federation of
United Nations Associations

Sir Crispin Tickell, Ambassador
Extraordinary and Plenipotentiary
Permanent Representative of the United
Kingdom to the United Nations

John Topping, President
Climate Institute

Eugene Tracy, Chairman of the Board
The National Energy Foundation

Garry Trudeau, Author

Stewart Udall, Author/Attorney

Randolph H. Ware, President
External Tanks Corporation

Alan N. Weeden, President
The Frank Weeden Foundation

James Whittaker
Mountaineer and Explorer

Paul Winter, Composer/Musician
Living Music

Senator Tim Wirth
State of Colorado

George M. Woodwell, President
Woods Hole Research Center

Peterson Zah, Director,
Western Regional Office
Save the Children Federation

This statement reflects the opinion of the individuals listed, but not necessarily that of
the institutions with which they are affiliated.

Valentin Kamenev, USSR Consul General to the United States, and Robert Redford

Photographs copyright © 1989 by John Schaefer

Stewart Udall, former United States Secretary of the Interior and IRM board member, with Paul Ehrlich, Bing Professor of Population Studies at Stanford University

Kakimbek G. Salykov, Member of the Congress of People's Deputies and Chairman of the Supreme Soviet Committee on Ecology and the Rational Use of Natural Resources, with Congresswoman Claudine Schneider.

Chief Oren Lyons, Faithkeeper of the Onondaga Nation

Terrell J. Minger, President of the IRM

Participants in Greenhouse/Glasnost: The Sundance Symposium on Global Climate Change, August 1989

Elvira Orlova, Senior Engineer at the Space Research Institute of the USSR Academy of Sciences; Roald Sagdeev, Member of the Congress of People's Deputies and former director of the Space Research Institute; Russell L. Schweikart, Apollo astronaut and President of the Association of Space Explorers; and Dr. Walter Orr Roberts, founder of the National Center for Atmospheric Research